Towards Better Work

Advances in Labour Studies

Advances in Labour Studies is a wide-ranging series of research titles from the International Labour Office (ILO), offering in-depth analysis of labour issues from a global perspective. The series has an interdisciplinary flavour that reflects the unique nature of labour studies, where economics, law, social policy and labour relations combine. Bringing together work from researchers from around the world, the series contributes new and challenging research and ideas that aim both to stimulate debate and inform policy.

Published in the series:

THE LABOUR MARKETS OF EMERGING ECONOMIES: HAS GROWTH TRANSLATED INTO MORE AND BETTER JOBS? (by Sandrine Cazes and Sher Verick)

BEYOND MACROECONOMIC STABILITY: STRUCTURAL TRANSFORMATION AND INCLUSIVE DEVELOPMENT (edited by Iyanatul Islam and David Kucera)

WAGE-LED GROWTH: AN EQUITABLE STRATEGY FOR ECONOMIC RECOVERY (edited by Marc Lavoie and Engelbert Stockhammer)

REGULATING FOR DECENT WORK: NEW DIRECTIONS IN LABOUR MARKET REGULATION (edited by Sangheon Lee and Deirdre McCann)

CREATIVE LABOUR REGULATION: INDETERMINACY AND PROTECTION IN AN UNCERTAIN WORLD (edited by Deirdre McCann, Sangheon Lee, Patrick Belser, Colin Fenwick, John Howe and Malte Luebker)

SHAPING GLOBAL INDUSTRIAL RELATIONS: THE IMPACT OF INTERNATIONAL FRAMEWORK AGREEMENTS (edited by Konstantinos Papadakis)

TOWARDS BETTER WORK: UNDERSTANDING LABOUR IN APPAREL GLOBAL VALUE CHAINS (edited by Arianna Rossi, Amy Luinstra and John Pickles)

Towards Better Work

Understanding Labour in Apparel Global Value Chains

Edited by

Arianna Rossi

Amy Luinstra

and

John Pickles

palgrave macmillan

First published 2014 by PALGRAVE MACMILLAN and the INTERNATIONAL LABOUR OFFICE

Palgrave Macmillan in the UK is an imprint of Macmillan Publishers Limited, registered in England, company number 785998, of Houndmills, Basingstoke, Hampshire RG21 6XS.

Palgrave Macmillan in the US is a division of St Martin's Press LLC, 175 Fifth Avenue, New York, NY 10010.

Palgrave Macmillan is the global academic imprint of the above companies and has companies and representatives throughout the world.

Palgrave® and Macmillan® are registered trademarks in the United States, the United Kingdom, Europe and other countries.

ISBN 978–1–137–37753–1
ILO ISBN 978–92–2–127802–3

This book is printed on paper suitable for recycling and made from fully managed and sustained forest sources. Logging, pulping and manufacturing processes are expected to conform to the environmental regulations of the country of origin.

A catalogue record for this book is available from the British Library.

A catalog record for this book is available from the Library of Congress.

Typeset by MPS Limited, Chennai, India.

Contents

List of Figures, Tables and Boxes

Figures

Tables

Boxes

Preface

The quality of jobs created by globalization of apparel production has long been under debate, and took centre stage following the tragic collapse of Rana Plaza on 24 April 2013 in which over 1,100 workers died and 2,500 were injured. The time is ripe for policy and institutional development which supports improvement in the conditions and rights of those who work in the global garment industry. This volume is intended to contribute to this goal by promoting greater understanding of the dynamics of labour in the industry's globalized value chain.

Better Work is a partnership of the International Labour Organization (ILO) and the International Finance Corporation (IFC) aimed at improving both working conditions and competitiveness in garment factories. The programme is committed to an evidence-based approach and has invested significantly in research and impact assessment including surveys of workers and managers to systematically measure the impact of its operations on workers and firms. Research informs our understanding of the effectiveness of our programme, how to improve it and where to focus scarce resources. By carefully measuring and communicating our impact we are held accountable to donors, businesses, and stakeholders with whom we work. Our research programme is also designed to influence public policy and business practices beyond the direct sphere of Better Work operations. With a solid knowledge base, well-analysed and peer-reviewed, we aim to build a much needed evidence base to support the case for the transformation in business, government and development policy and practice that is required to drive sustained improvement in job quality.

The contributions collected in this edited volume were first presented at a conference convened by Better Work in 2011, 'Workers, Businesses and Government: Understanding Labour Compliance in Global Supply Chains'. The knowledge generated and shared at the conference by the leading experts in this field advanced our understanding of the assumptions upon which we work, such as the economic viability and business case for labour standards compliance, as well as the positive impact for workers and their families that improving working conditions can have. Evidence from a broad spectrum of case studies, regions and countries allows us to have concrete examples of what works and what can be improved, for Better Work, for governments and for the stakeholders in the global garment industry.

The contributors to this book and their methodological approaches are diverse by design. Despite this the volume develops clear conclusions that have informed the strategy that ILO and IFC have developed for Better Work in the years to come. The findings suggest that the effectiveness of initiatives aimed at improving working conditions, whether private or public, are constrained by the fragmented structure of global garment production. Efforts of lead firms, trade unions, civil society groups, and national or international agencies to improve the conditions of work must also be supported by broader trade and development policy which paradoxically often drives fragmentation and undermines opportunities for social upgrading and stronger governance of the labour market. A key pillar of our approach in the next three years will be to inform our institutional partners' advocacy and interventions in this area.

This book marks the end of the beginning of our research programme. In the years ahead we will continue to share the rapidly growing data, lessons and evidence that arise from our own work and our collaboration with others.

Dan Rees
Director, ILO-IFC Better Work Programme

This book is published at a critical time for workers employed in export industries. There is growing recognition, including by the World Trade Organization (WTO), the World Bank, the Organisation for Economic Co-operation and Development (OECD) and United Nations Conference on Trade and Development (UNCTAD), that trade increasingly takes place through global value chains (GVCs). Within GVCs lead firms (brands, retailers, large manufacturers and traders) coordinate supply linking final consumers with their networks of suppliers. Outsourcing generates extensive employment, particularly in developing countries. However, the extent to which workers benefit from participation in GVCs remains an open question. Global production networks provide new income earning opportunities, particularly for women workers. However, conditions are often poor, and the tragedy of the 2013 Rana Plaza factory collapse in Bangladesh demonstrated the risks some workers face. A better understanding of both the benefits and challenges of participating in GVCs is essential for improving the conditions and rights of workers.

Capturing the Gains is an international research network investigating linkages between the commercial dynamics and outcomes for workers within GVCs. Capturing the Gains research draws together two concepts: (a) economic upgrading, defined as moving to higher value activities by firms; (b) and social upgrading, broadly defined as decent work and enabling rights for workers. Our comparative research has focused on four sectors: apparel, agrofood, tourism and mobile communications across countries in Africa, Asia and Latin America. A key finding of the research is that economic upgrading of firms *can* but does *not necessarily* lead to social upgrading of workers. Outcomes vary for different groups of workers and by context (even in the same sector), with downgrading a reality for many firms and workers. How GVCs are governed and relative bargaining positions of workers are critical determinants in addressing the question of who has better jobs.

This volume brings together a unique collection of papers from the ILO/IFC Better Work programme conference held in Washington, DC, in 2011. Many chapters present findings from Capturing the Gains research on apparel, complementing studies linked to the Better Work programme. This novel collection provides important insights into the complexities and challenges of improving workers' conditions and rights in global apparel production. Importantly, it contributes to seeking more effective private, public and civil society strategies and policies to ensure *all* workers – who are the pillars of apparel production – enjoy greater benefits from participating in this global industry.

Stephanie Barrientos, University of Manchester
Gary Gereffi, Duke University

Notes on Contributors

Dennis Arnold is Assistant Professor of Globalization and Development at Maastricht University, the Netherlands. He publishes and teaches on three interrelated areas: labour, migration and citizenship; global production network analysis; and borderlands of continental South-East Asia. He does research in Cambodia, Myanmar, Thailand and Viet Nam. His work has appeared in *Antipode, Geography Compass, Journal of Contemporary Asia, American Behavioral Scientist,* Routledge edited books and a monograph published by Mahidol University's Human Rights in Asia Book Series.

Jennifer Bair is Associate Professor of Sociology at the University of Colorado at Boulder, where she is also a faculty affiliate in the Institutions programme at the Institute for Behavioral Science. Her research in the political economy of gender, globalization and development has appeared in journals such as *World Development, Global Networks, Economy and Society, Social Problems* and *Signs*. She is the editor of *Frontiers of Commodity Chains Research* (2009) and co-editor of *Workers Rights and Labor Compliance in Global Supply Chains* (2013) and *Free Trade and Uneven Development: The North American Apparel Industry after NAFTA* (2002).

Thomas Bernhardt is a PhD candidate in Economics at the New School for Social Research, United States. Before joining the New School as a Fulbright Scholar, he studied Economics as well as International Business at the Vienna University of Economics and Business, and worked as junior fellow at the Austrian Institute of Economic Research (WIFO) and as trainee at the Brussels Office of the Austrian Chamber of Labour (AK EUROPA). He currently works as a consultant for the United Nations Industrial Development Organization (UNIDO). His research interests focus on trade and economic development, global value chains and economic policies for development.

Resmi P. Bhaskaran is a development economist with 16 years of experience in social and public policy research and advocacy. Her focus is on urbanization, women, child rights, migration and microfinance. She has published in Indian and international journals. She has worked with Sa-Dhan (microfinance association in India), Institute for Human

Development and Save the Children. Currently, she is working as a freelancer and serving as research advisor to National Commission of Protection of Child Rights and National Right to Education Forum in India. Her research received an award from the South Asian Network of Economic Research Institute of GDN in 2008 at the South Asia regional level.

Drusilla Brown is Associate Professor of Economics and Director of the International Relations Program at Tufts University, United States. She received a PhD in Economics from the University of Michigan in 1984. Her expertise includes the use of a large scale computable general equilibrium model to analyse the economic effects of international trade agreements. More recently, her research has focused on issues concerning the impact of globalization on wages, employment and working conditions, including child labour. Currently, she is a member of the interdisciplinary team undertaking monitoring and evaluation of Better Work. Recent publications include *Globalization, Wages, and the Quality of Jobs: Five Country Studies* (with R. Robertson, G. Pierre and M. Sanchez-Puerta, 2009) and articles in the *Review of Development Economics, World Development* and the *Journal of Economic Perspectives.*

Upendranadh Choragudi is currently working as Programme Support Coordinator of ActionAid International working on issues of public services and redistributive public policies. Previously, when this research was being carried out, he worked as Senior Fellow at the Institute for Human Development, New Delhi.

Rajeev Dehejia is Professor of Public Policy in the Robert F. Wagner Graduate School of Public Service at New York University. He received his PhD in economics from Harvard University in 1997. His research spans econometrics, development economics, labour economics and public economics, with a focus on empirical microeconomic policy research. He is a Faculty Research Fellow of NBER, a Research Fellow at IZA and a Research Network Fellow at CESifo.

Gary Gereffi is Professor of Sociology and Director of the Center on Globalization, Governance and Competitiveness at Duke University. He received his BA from the University of Notre Dame and his PhD from Yale University. Gereffi has published numerous books and articles, including *The New Offshoring of Jobs and Global Development* (2006) and *Global Value Chains in a Postcrisis World: A Development Perspective* (2010). He has recently completed the three-year 'Capturing the Gains' research project, and he is currently working on global value chains in emerging economies and new methodologies for measuring value chain upgrading.

Shane Godfrey is Senior Researcher at the Labour and Enterprise Policy Research Group, University of Cape Town, South Africa. His research interests are the organization of work and labour market regulation, including the implications for economic development of different contractual and regulatory regimes. He has a special interest in labour-intensive consumer goods manufacturing sectors and the relationship of manufacturing firms with retailers. He has published widely on labour market regulation in South Africa.

Stephanie Kuttner is a Gender and Social Development Consultant based in Brasilia. She coordinated the World Bank's recent study on gender and labour conditions in the Lao garments sector, including firm surveys with factory managers and focus group discussions with workers. She is a contributor to the 2012 Lao Country Gender Assessment for the World Bank and Asian Development Bank, and to various other governance and social development assessments in Brazil, Lao People's Democratic Republic, Haiti and countries in the Central African region. Stephanie holds a PhD and an MPhil in International Relations from the University of Oxford.

Amy Luinstra is a Senior Program Officer in the IFC's Sustainable Business Advisory department where she primarily works on the Better Work programme. She serves on the management team of Better Work and leads the team responsible for policy and research. She was on secondment from the World Bank to the ILO to help launch Better Work in 2007 through early 2011. Prior to Better Work, she was a Social Protection Specialist in the World Bank Social Protection and Labor unit, where she worked on labour market policy and coordinated the Bank's dialogue with the international trade union movement.

Frederick Mayer is Associate Professor of Public Policy and Political Science and Director of the Program on Global Policy and Governance at Duke University's Sanford School of Public Policy. His research focuses on globalization and its effects, with particular emphasis on the labour and environmental effects of economic integration. Mayer received a BA in history and literature from Harvard College, and an MPP and a PhD in public policy from the John F. Kennedy School of Government at Harvard University.

Doug Miller is Emeritus Professor, Worker Rights in Fashion, University of Northumbria, United Kingdom. Between 2000 and 2008 he was seconded from the university as research director at the International Textile Garment and Leather Workers' Federation, the global union for the sector. In 2008 he returned to an academic post in the Design School sponsored by the global union and the Spanish retail fashion multinational Inditex SA, where he was responsible for developing teaching and research in the area of 'ethical fashion'.

Dev Nathan is a Professor at the Institute for Human Development, New Delhi and a Visiting Research Fellow at the Center on Globalization, Governance and Competitiveness, Duke University, United States. He has been one of the research coordinators of the Capturing the Gains study of global value chains. Currently, he is Coordinator of the South Asia Labour, Employment and Social Protection Programme being implemented by the Institute of Human Development in collaboration with the United Nations Economic and Social Commission for Asia and the Pacific and the ILO Office for South Asia in New Delhi. Besides labour conditions in global production, his research interests include development issues of indigenous peoples. He recently co-authored *Markets and Indigenous Peoples in Asia* (2012).

Nicola Phillips is Professor of Political Economy at the University of Sheffield, United Kingdom. Her research focuses on global political economy and global development, with specific interests in labour in global production, unfree labour and human trafficking, and migration and development. Her recent works include *Development* (co-authored with Anthony Payne 2010), *Migration in the Global Political Economy* (edited volume, Lynne Rienner 2011), and articles in the *Review of International Political Economy*, *Economy and Society* and *Global Networks*. She currently holds a Major Research Fellowship from The Leverhulme Trust, for research on forced labour and human trafficking for labour exploitation in the global economy.

Kabmanivanh Phouxay is currently Head of the Research and Postgraduate Division, Faculty of Social Sciences, National University, Lao People's Democratic Republic. She has worked with the Mekong Migration Network since 2003.

John Pickles is the Earl N. Phillips Distinguished Professor of International Studies in the Department of Geography at the University of North Carolina at Chapel Hill. He received his BA degree from Oxford University and holds PhDs from the University of Natal and the Pennsylvania State University. He has published numerous books and articles, including: *Globalization and Regionalization in Post-Socialist*

Economies (2009), *State and Society in Post-Socialist Economies* (2008) and *A History of Spaces: Cartographic Reason, Mapping and the Geo-Coded World* (2004). He directed the apparel research for Capturing the Gains and is currently working on European global production networks and Euro-Med migration and border policies.

Kelly Pike is a Postdoctoral Fellow at York University in Toronto, Canada. She recently received her PhD from Cornell University, United States. Her dissertation *Made in Lesotho: Examining variation in clothing workers' perceptions of labour standards compliance* was informed by two years of field work in southern Africa, during which time she also worked with Capturing the Gains and Better Work Lesotho. Her research interests focus on labour standards and forms of social regulation in global value chains.

Leonhard Plank is a business economist and works at the Centre of Public Finance and Infrastructure Policy at the Department of Spatial Development, Infrastructure and Environmental Planning at the Vienna University of Technology. He received his Doctorate in Business Administration at the University of Graz, Austria. His research and work focuses on transnational corporations, global production networks and regional development.

Richard Record is a Senior Economist in the World Bank's South-East Asia Poverty Reduction and Economic Management team. Based in Vientiane, Lao People's Democratic Republic, he manages the World Bank's country programme on trade and competitiveness including project financing to support automation of customs and border systems, trade policy and negotiations capacity-building, and competitiveness interventions aiming to improve productivity in the Lao garments manufacturing sector through fee-based training services and a factory standards improvement scheme based on the Better Work model.

Raymond Robertson is Professor of Economics at Macalester College, United States. His research focuses on the union of international, labour and development economics. He has published in *American Economic Review, Review of Economics and Statistics, Journal of International Economics, Review of International Economics, Journal of Development Economics* and other journals. He is a non-resident fellow at the Center for Global Development and a member of the State Department's Advisory Committee on International Economic Policy. He received his PhD from the University of Texas after spending a year in Mexico as a Fulbright Scholar.

Arianna Rossi is the Research and Policy Officer for the ILO/IFC Better Work programme. Prior to that, she was Research Officer for the Capturing the Gains research network at the University of Manchester, United Kingdom. Her research and work focus on labour rights in global production networks, corporate social responsibility, gender and the impact of multi-stakeholder initiatives on workers' wellbeing and firms' competitiveness. She received her PhD from the Institute of Development Studies at Sussex University with a dissertation on social upgrading in global production networks, analysing the garment industry in Morocco.

Cornelia Staritz is Senior Researcher at the Austrian Research Foundation for International Development (ÖFSE). Prior to that, she worked at the International Trade Department of the World Bank and at the Vienna University of Economics and Business. She holds Master degrees in Economics and Commerce, a Doctorate in Economics from the Vienna University of Economics and Business and a PhD in Economics from the New School for Social Research. Her research focuses on economic development, international trade, global value chains and production networks, private sector development and commodity-based development.

Daniel Vaughan-Whitehead is Senior Specialist at the International Labour Office, where he is responsible for wages and working conditions around the world. In 2008, he initiated the ILO's *Global Wage Report*, which has become established as a worldwide reference in this area. He is also a Professor at Sciences Po in Paris, and is the author of many books and articles on wages, industrial relations, social dumping, forms of workers' participation and social policies in general. Previously, he worked as adviser to European Commission President Jacques Delors and for the European Commission, where he was responsible for social dialogue in the EU enlargement process.

List of Abbreviations

AMV	actual minute value
AGOA	African Growth and Opportunity Act
BFC	Better Factories Cambodia
CAFTA	Central America Free Trade Agreement
CAFTA-DR	Dominican Republic–Central America Free Trade Agreement
CCAWDU	Coalition of Cambodian Apparel Workers Democratic Union
CCT	conditional cash transfer
CM	cut and make
CMT	cut, make and trim
CPP	Cambodian People's Party
CSR	corporate social responsibility
CNZF	National Free Zones Commission (Nicaragua)
EPZ	export processing zone
EU	European Union
FDC	fixed duration contract
FGDs	focus group discussions
FLA	Fair Labor Association
FOB	freight on board
FTA	free trade agreement
GMAC	Garment Manufacturers' Association of Cambodia
GPN	global production network
GSCP	Global Social Compliance Programme
GSD	General Sewing Data
GVC	global value chains
HR	human resources
IFA	international framework agreement

IFC	International Finance Corporation
ILO	International Labour Organization
IMF	International Monetary Fund
INATEC	National Institute of Technology (Nicaragua)
ITGLWF	International Textile, Garment and Leather Workers Federation
MFA	Multi-Fiber Arrangement
NAFTA	North American Free Trade Agreement
NCR	National Capital Region (India)
NGO	non-governmental organization
OBM	own brand manufacturer
ODM	own design manufacturer
OECD	Organisation for Economic Co-operation and Development
OEM	own equipment manufacturer
OPT	outward-processing trade
OSH	occupational safety and health
PICCs	Performance Improvement Consultative Committees
PTS	pre-determined time standard
QIZs	qualified industrial zones
RMB	Chinese renminbi
SAM	standard allowed minute
SKU	stock-keeping unit
SMV	standard minute value
TPLs	tariff preference levels
UDCs	undetermined duration contracts
USAID	United States Agency for International Development
WRAP	Worldwide Responsible Accredited Production
WRC	Workers Rights Consortium
WTO	World Trade Organization

Introduction

Arianna Rossi, Amy Luinstra and John Pickles

The manufacture of apparel is, in many ways, an exemplar of global production. Since the 1970s, multinational brands have increasingly outsourced their manufacturing activities to lower cost production locations in developing countries. The international trading regime then regulated by the Multi-Fibre Arrangement (MFA) contributed to the development of apparel industries in a large number of developing countries. The low entry barriers and minimal investments needed in apparel led to booming employment in apparel factories in regions where formal employment was limited or, in some cases, entirely absent. New opportunities were created especially for young, unskilled women and migrant workers who had access to waged labour for the first time.

While this translated into higher labour force participation rates and new empowerment opportunities for these previously marginalized groups, it also appeared increasingly clear that the globalization of apparel export production has been one of the major triggers of poor working conditions and a significant cause of regional wage depression. With increasing fragmentation of sourcing, heightened competition among buyers, and the rise of large buyer-driven value chains, contract prices were squeezed and poorly regulated second- and third-tier subcontracting became increasingly common. Intensification of competition, weakening industry and governmental regulation, and expanded volatility and uncertainty in markets have combined to create new and important dynamics in the organization of global apparel industries.

In 2005, quota removal expanded the range of opportunities for footloose sourcing, which in turn expanded employment opportunities in some regions, but often at the cost of more widespread predatory employment practices, feminization of work and depressed wages. As a consequence, labour issues and violations of labour standards in global apparel production have become endemic and many argue that the global apparel production network is built on a race to the bottom, squeezing labour rights and standards in the process.

This volume seeks to provide a new analytical perspective on these issues, by looking at the role of different global and local actors and at why global value chain (GVC) dynamics have the potential to either promote or hamper compliance with labour standards and worker wellbeing.

To what extent do global value chains contribute to improving social welfare of workers and their communities, and to what extent are they exacerbating the problems of low-wage, highly mobile, and exploitative working conditions?

The apparel global value chain

The emergence of global coordinated production as a key feature of economic globalization has been studied extensively in recent decades from different disciplinary perspectives, ranging from the historically- and macro-oriented World Systems commodity chains (Hopkins and Wallerstein 1977) to the more operational and firm-centred global commodity chains, global value chains (e.g. Gereffi 1994, 1999; Kaplinsky 2000; Humphrey and Schmitz 2002) and the broader concept of global production networks (Henderson et al. 2002; Hess and Coe 2006; Coe et al. 2008).

The initial conception of global commodity chains (Gereffi and Korzeniewicz 1994) focused on the linkages between the production, distribution and consumption of commodities. It described how they are globally interconnected along chains that embody networks of activities and actors, and where the power to produce and capture value occurred along the chain. This early GVC analysis did not include considerations related to the scale and quality of employment and the broader contexts in which chains are embedded (a notable exception is Gereffi 2006). Henderson et al. (2002) addressed this challenge by focusing on global production networks and their embeddedness in territorial production networks and social and institutional contexts. This approach stems from the recognition that economic activity and actors' behaviours are strongly influenced by the social context in which they operate (Granovetter 1985), and encompasses not only the economic and commercial actors involved in global production, but also the whole range of actors operating in the social and institutional context that surrounds and influences global production such as states, supra-national organizations, business associations, trade unions and non-governmental organizations (NGOs).

The apparel value chain is organized around five main segments (Figure I.1): raw material supply, input supply, manufacturers (including their domestic and overseas subcontractors, as well as embellishers); export channels, sometimes managed by intermediary trading companies (such as Li & Fung) and always organized through logistics freight forwarding companies; and marketing and retail.

Over time, the national structure of manufacturer-driven value chains has given way to increasingly fragmented production systems in globalized buyer-driven and retailer-driven value chains (Gereffi and Frederick 2010; Gereffi and Memedovic 2003: 5). For Fernandez-Stark et al. (2011: 7) the apparel industry is 'the quintessential example of a buyer-driven commodity chain marked by power asymmetries between the suppliers and global buyers of final apparel products.' In the process, some lead firms that control design, branding and marketing have been able to exercise strong control over sourcing decisions, and hence over how, when and where specific parts of the production process will take place. In so doing, lead firms have been able to control where value is extracted and to whom profit accrues at each stage, essentially determining how basic value-adding activities are distributed along the value chain (Fernandez-Stark et al. 2011).

In this process, value in globalized value chains has increasingly been captured by input suppliers (primarily yarn and fabric), up-front and end-market services, research, development, design, marketing and retail services. Actual assembly operations (primarily stitching and embellishment) and logistics costs have been squeezed, and the main actors have had little positional or negotiating power vis-à-vis the lead firms (Frederick and Gereffi 2011). This fragmentation and globalization of each segment has prevented backward and forward linkages emerging in many low income countries. As a result, returns to capital have increased while returns to wages have generally declined. In this process, lead firms (and increasingly retailers and network organizers like Li & Fung) have played progressively important coordination roles in managing production and delivery, and in turn have been able to capture a larger proportion of total chain value.

As Bair (2005) notes, the earlier focus of much of this work on global commodity chains has more recently shifted towards analysis of the way that value chains are organized and governed, and a consideration of the implications for industrial/economic upgrading (see Gereffi et al. 2005; Sturgeon 2009). In this later literature a primary focus has been

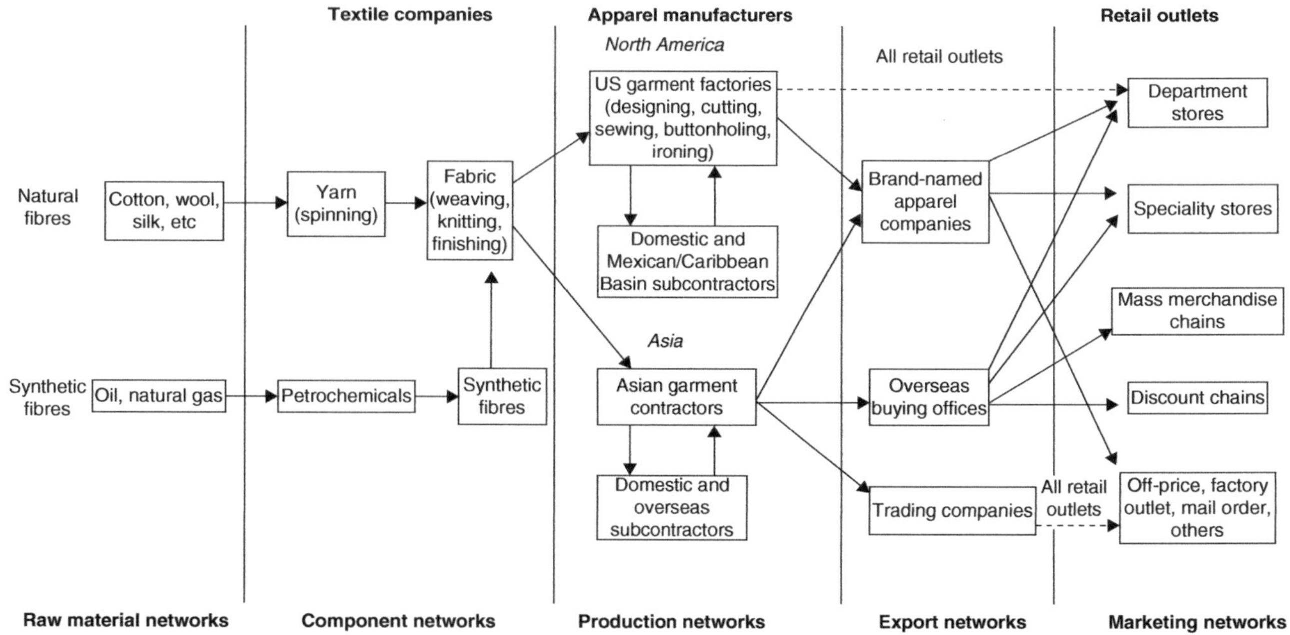

Figure I.1 Apparel production system
Source: Gereffi and Memedovic (2003).

on the mechanisms whereby firms and industries engineer a process of economic upgrading within global value chains to capture additional functions in supply chains, which generate higher value added. Humphrey and Schmitz (2002), for example, distinguish between four types of upgrading in global value chains: product, process, functional and chain upgrading.

- *Product upgrading* refers to new configurations of product mix towards higher value added products. The manufacturer's enhanced capacity enables it to produce more complex products requiring higher levels of skill, more technical processes and specialized knowledge.
- *Process upgrading* involves firms retaining their position in a chain by enhancing productivity gains through adopting new production processes, and usually requires new investment and a focus on skill development as new machines or line operations are changed, such as in lean manufacturing.
- *Functional upgrading* involves a movement 'up' the chain into newer, higher value added activity, such as full package and own design/own brand manufacturing in the clothing sector. Four main categories characterize apparel manufacture:

    1. *CM/CMT*: 'cut and make' and 'cut-make-trim' are the two most basic forms of apparel production, particularly in offshore contracts. Suppliers are contracted to stitch-up fabric and other inputs according to the specifications provided by the buyer, limiting operations to cutting, sewing, embellishment and trim, and shipping the ready-made garment. CM/CMT contracts have typically been common in export processing and free-trade zones, and have usually been driven by outward processing trade agreements and preferential tariffs.
    2. *OEM*: own equipment manufacturers offer a wider range of production capacities and services to buyers, including limited design, warehousing and embellishment. They may also be responsible for sourcing upstream inputs (fabric, dyeing and trim) either from designated suppliers or from their own suppliers. The supplier also assumes responsibility for some part of the logistics chain for the finished orders. With added capacities, own equipment manufacturers may become full package suppliers, carrying out the entire production process for a given order.
    3. *ODM:* own design manufacturers carry out all parts of the production process, including the crucial design functions that enable much greater control over ordering and timing of input supplies,

development of new lines and construction of samples, selecting fabric designs and hence managing the timing and costs of production much more directly. Full package suppliers may also handle delivery to the final customer, as well as build independent labels for domestic markets.
4. *OBM:* own brand manufacturer further expands the capacities of the manufacturer and requires the addition of research and development, design and marketing functions. Successful OBM manufacturers may be able to leverage their domestic market labels into regional or global markets.

- *Chain upgrading* involves leveraging expertise gained in one industrial sector to enter a new sector, which may imply higher skills and capital requirement and value added.

In the context of GVC research, the analytical focus has shifted from an earlier emphasis on the significance for economic development of the difference between buyer-driven and producer-driven commodity chains (Gereffi 1994) to one investigating the developmental implications of upgrading, including on labour and social outcomes (Bair 2009; Barrientos et al. 2011; Cattaneo et al. 2010; Rossi 2013a; Tokatli 2007). As Plank and Staritz (2009: 66) have argued, attention is required beyond the black box of the firm to consider also who benefits from upgrading:

> Even if firms gain rewards for their upgrading efforts, the rewards may not be passed on to workers in the form of higher wages, greater job security or improved working conditions. Firm upgrading may even be based on deteriorating working conditions.

With hindsight, it was perhaps inevitable that global sourcing and value chain fragmentation would lead to loss of tight control over production and working conditions. As production was outsourced to 'stitch-up' shops, work was deskilled and assembly firms were separated organizationally and geographically from other value-generating parts of the value chain (design, yarns, fabrics, dyeing and marketing). Suppliers and their workers have, as a result, become the weakest actors in GVCs, increasingly trapped in conditions of input and order dependency and subject to footloose sourcing practices. As secondary and tertiary subcontracting expanded across many supplier networks in different countries, workplace conditions deteriorated. In this context, there are very limited opportunities for economic and especially social upgrading.

Labour in global production

While the effects of global production on working conditions in supplier networks was central to the development of GVC analyses of economic upgrading and competitiveness, it remained largely implicit and under-theorized (Bair 2009). Particularly in its tendency to value economic upgrading as a model for business and national competitiveness, other questions of job loss, low-wage employment and deteriorating working conditions were not addressed sufficiently (Pickles and Smith 2010; Bair and Werner 2011). Some have argued, as Selwyn (2013) does, that GVC theories have yet to account for the relationship between exploitation resulting from weak governance and the more systemic exploitation involved in all low-wage work.

One attempt to redress this lacuna is the growing body of literature from academia and from civil society actors dealing with the role of labour rights and working conditions in global value chains (Barrientos et al. 2011; Rainnie et al. 2011; Selwyn 2011). This literature introduces the concept of social upgrading, that is, the process of improvements in the rights and entitlements of workers as social actors by enhancing the quality of their employment (Barrientos et al. 2011). This concept has been operationalized in different ways. At the macro level, studies (e.g. Milberg and Winkler, 2011) have defined social upgrading as growth in employment and real wages. This is the same 'parsimonious approach' used by Bernhardt in Chapter 2 of this volume. In other cases, the definition of social upgrading has made use of the ILO's decent work agenda and focuses on two main issues: *measurable standards*, including wages, physical wellbeing (including working hours and occupational safety and health) and employment security; and *enabling rights*, such as freedom of association, voice and non-discrimination.

Despite the growing attention to the issue of social upgrading, the governance mechanisms aimed at sustaining these processes are still understudied. Mayer and Posthuma (2012) identify different forms of governance relevant to global production networks (GPN) analysis: *public governance*, comprising of national government policies and international policies such as trade policies and international labour standards; *private governance* led by private sector actors; and *social governance* related to worker organizations and civil society. So far, most of the GVC and GPN literature looking at labour and working conditions have zeroed in on private governance initiatives, while labour studies and sociology of work has tended to focus more on the role of state regulation. Indeed, research focusing on private governance initially focused on the role of 'lead firms' in GVCs as actors not only responsible for decision making in production, but also for the working conditions in the factories manufacturing their products in the developing world. In particular, corporate social responsibility (CSR), buyers' codes of conduct, their content and their impact on working conditions and labour rights became more central to GVC analyses (see e.g. Mamic 2003; O'Rourke 2002; Barrientos and Smith 2007; Locke 2013). Considerably less attention has been devoted to the role of public governance and in particular to the need of effective public enforcement of labour law. The recent changes in the apparel industry have led to a new consensus that private governance alone cannot address the complexities of industrial relations in global value chains and – by extension – that the responsibilities and costs of social upgrading must be shared, with the state assuming a greater role in the provision of essential common pool resources (such as energy and logistics infrastructure, workforce development, health and safety regulations, and labour dispute mediation). Pushing this argument further, it is increasingly believed that combination of public and private interventions is critical to achieve sustainable change in working conditions in GVCs (Braithwaite 2002; Kolben 2007; Locke et al. 2012). Also, in terms of social governance, civil society and advocacy, organizations have focused their research efforts on explorations of factory workers' lives, opportunities and challenges as a result of their participation in GVCs (see e.g. Brooks 2007; Hale and Wills 2006; Oxfam International 2004; Rivoli 2006). More recently, increasing attention is being given to civil society actors and their role in driving social upgrading and ethical sourcing initiatives (Bair et al. 2013).

While international buyers, supplier firms and workers are crucial actors concerned with labour issues in GVCs they are not the only ones. In order to broaden the range of actors and the web of relationships that influence outcomes for firms and workers, the chapters in this book include social, institutional and political actors operating at global and local levels in GPNs, underlining the need to look more holistically at governance actors to fully understand the opportunities for economic and social upgrading in GPNs. This book makes three main interventions. First, the book focuses on labour from a governance and policy perspective highlighting the role of this broader set of actors involved in the global apparel value chain. Second, the collection analyses the extent to which these policy and institutional actors have been successful in improving the conditions of work and enabling rights of workers in apparel value chains. And third, the book seeks to add to these analyses important assessments of the extent to which the social upgrading of workplace conditions and worker rights in apparel value chains may have positive impacts on the economic conditions of specific firms, regions and chains — in other words — whether there is a business case for social upgrading. The findings in this volume suggest that the effectiveness of initiatives aimed at improving working conditions – whether private or public – is constrained by the globalized and fragmented structure of GVC production. Private initiatives by lead firms and public initiatives by civil society groups and national or international agencies to improve the conditions of work can only occur and be sustained within the broader determining context of international and national trade policy, much of which produces and/or reinforces the fragmented structure of GVC production and sustains systemic limits to what social upgrading can be achieved.

Overview of chapters

This book explores different experiences and initiatives around better working conditions seeking to understand the role and perspectives of different actors in GPNs in addressing labour. The chapters focus specifically on governance mechanisms in GVCs and provide solutions-oriented approaches for promoting improved working conditions and labour rights in the apparel industry. Chapters analyse how in some contexts workers, governments and business are collaborating to confront some of the key opportunities and challenges pertaining to labour in global apparel value chains, and how in other contexts they are dealing with the obstacles that can arise in such collaboration. Most of the chapters are based on detailed and grounded case studies of specific country cases by researchers who have been actively studying apparel value chains and conditions of work for many years. All address in particular ways the reciprocal effects of economic and social upgrading on the conditions of work in apparel production.

This collection of chapters is articulated in two main parts. Part I reflects on global governance, economic and social upgrading, and workers' wages in GPNs, setting out the analytical framework as well as the broad context (both in terms of institutions and of trade) where global apparel production is situated. In Chapter 1, Mayer and Pickles argue that parallel to an increase in trade that expanded investment and employment opportunities, and increased aggregate returns to capital, there has been growing global inequality, deepening exploitation and casualization of workers at the margins of the new global economy and pressure on labour in more traditional centres of production. Mayer and Pickles have found that lead firms, suppliers, states and civil society actors are responding to this global 'governance deficit' (Gereffi and Mayer 2006) in a wide range of ways to promote decent work in apparel global value chains. They define governance as those institutions that constrain or enable market actor behaviour, both public in the form of governmental policies, rules and regulations, and private in the form of social norms, codes of conduct adopted by businesses, consumer demand for social responsibility or other non-governmental institutions and social movements.

Complementing the analytical framework provided by these contributions, Bernhardt, in Chapter 2, analyses economic and social upgrading in apparel global value chains from a trade and employment perspective at the macro level. In particular, he explores the hypothesis that economic upgrading leads to social upgrading: economic theory generally supports the view that upgrading translates quickly and fully into social upgrading, including improved wages and labour standards. Using a 'parsimonious approach' in defining economic and social upgrading, Bernhardt combines these measures to analyse how they are related to each other in key apparel producing countries, highlighting both the cases in which the two measures go hand in hand, and those for which evidence is mixed. The chapter concludes by discussing the significance of 'downgrading' in specific countries and analysing the causal link between the economic and social realms.

Following these chapters dealing with governance, institutions, trade and employment from a macro perspective, the book turns to a key component of social upgrading: wages and wage practices along the apparel GVC. In Chapter 3, Vaughan-Whitehead tackles this issue of paramount importance to workers' livelihoods and analyses how 'fair' wage practices are along the supply chain. He defines a 'fair wage approach' that aims to provide CSR actors with a coherent set of fair wage dimensions and indicators. The chapter analyses wages in more than 100 suppliers in Asia and is complemented by three case studies in China in 2010–11. The results provide first-hand information on wage practices among suppliers and identify a number of wage problems along the supply chain. The findings suggest that there is an urgent need to address wage issues using a broad spectrum of 'fair wage' dimensions, including living wages, minimum wages, prevailing wages, social dialogue, the payment of working hours and the evolution of wages in accordance with prices, enterprise performance and changes in technology and human capital.

Complementing Vaughan-Whitehead's analysis, in Chapter 4 Miller draws on time systems methodology and develops a model by which buyers can assist vendors to raise pay in line with a multinational's 'living wage' code commitment where such exists. Using an example of a five-pocket western style jean made in Cambodia, and factory wage data, he calculates the differential between a prevailing and living wage cost on a given order and outlines ways in which this supplement can be delivered to a manufacturing facility and passed on to workers. 'Sustainable' labour costing using this method has implications for buying practices and for industrial relations at the factory level, since information disclosure, worker organization and capacity building are key to delivery. The chapter concludes by arguing that the success of living wage campaigns will ultimately depend on the extent to which buyers and suppliers are committed to the core principles of freedom of association and collective bargaining.

Part II of the volume brings the focus to the country level, examining particular national contexts and initiatives that offer potential for improved governance and upgrading. The chapters in this part highlight how different national and international contexts support or challenge the attainment of social upgrading in apparel producing supplier countries.

Through case studies of Morocco and Romania in Chapter 5, Plank, Rossi and Staritz focus on the effects of 'fast fashion' on creating jobs characterized by a high degree of flexibility, uncertainty and precariousness. Fast fashion adds new requirements of ever shorter lead times, improved responsiveness and greater flexibility to existing cost and high quality demands. The effects on workers' prospects for social upgrading are not clear. For many workers, the pressure for quicker delivery time results in the downgrading of working conditions. However, where fast fashion practices are mediated by local institutional structures and regulatory contexts, social upgrading opportunities may be possible. In their case studies, the authors show how Moroccan and Romanian firms reacted in different ways to accommodate these pressures. In both cases, flexibilization of work contracts increased and the co-existence of regular and irregular workers became the normal way to cope with conflicting pressures of low costs, high flexibility, high quality, consistency and reliability. The result has been the emergence of bifurcated labour conditions and differential labour rights and relations within factories. Their findings pose serious questions for analyses of social (and economic) upgrading that work with aggregate firm-level or industry-level data, and fail to disaggregate the effects of industrial dynamics by type of worker. Furthermore, the authors identify different governance measures (an industry-led social compliance initiative in Morocco and public governance in Romania) as critical factors mediating the pressures created by fast fashion.

In Chapter 6, Record, Kuttner and Phouxay assess the consequences for working conditions and productivity of high transport costs, low labour productivity and high staff turnover rates in the Lao garment sector. While the authors find relatively few reported instances of extreme violations of international labour standards, workers – most of whom are female migrants aged 16–25 – suffer from frequent compulsory overtime and difficult working environments, and have a very limited understanding of their contractual rights and obligations. In this context, they tend to see work in the garment sector as a temporary phase in their life during which their primary goal is to earn income to subsidize future activities, such as sibling education, having their own family or starting their own small business. Perspectives on working conditions differ dramatically between management and workers, with frequent misunderstandings and cultural clashes on the factory floor. When faced with recurring frustrations and limited formal representation, workers are left with little option other than the withdrawal of participation in the garments labour market. Firms struggle to raise productivity when faced with such high levels of worker turnover and skills leakage. The chapter concludes with several indicative recommendations, building both a 'business case' and a 'development case' for how the Lao garment sector can break out of this suboptimal and unsustainable position.

In Chapter 7, Bhaskaran, Nathan, Phillips and Upendranadh analyse the changing conditions of child labour in segments of the apparel value chain in India's National Capital Region. The chapter shows that child labour exists almost entirely in the embroidery and embellishment tasks of garment production, either as part of homework or in household-based enterprises. After presenting empirical findings, the chapter considers various aspects of the child labour question, including the roles of poverty and lack of a capacity to aspire as drivers for child labour. The chapter considers possible governance options to eliminate child labour, including the role of the new constitutional amendment for universal and compulsory education and a possible conditional cash transfer scheme.

In Chapter 8, Pike and Godfrey use baseline Better Work data and firm-level and worker interviews from Lesotho to analyse the factors that lead to variation in labour standards compliance in the apparel industry in

South Africa and Lesotho. The authors argue that ownership, end-user markets and dispute resolution practices play crucial roles. The work of activists, NGOs and global union campaigns has been a key force in pressuring multinational companies to be responsible for working conditions in their supplier factories. Traditionally, this has translated into codes of conduct that brands ask their suppliers to adhere to as a condition of their continued business. While this has pressured big name brands to be responsive and manufacturers to be more alert, it has not always had the intended effect of improving conditions for workers. In addition to the challenges of monitoring these codes, rigid enforcement of labour standards can lead to workers losing their jobs if brands discontinue contracts with noncompliant suppliers. An alternative to the traditional channel for labour standards enforcement is a multi-stakeholder approach, which directly engages retailers with unions and manufacturers, domestic government and NGOs. This kind of social dialogue facilitates greater transparency, accountability and incentive for stakeholders to forge opportunities for actual improvements in labour standards.

In Chapter 9, Arnold focuses on the regulation of industrial relations in Cambodia's textile and garment industry – a unique ensemble of state, trade union, private sector and international institutions that is promoted as a 'fair model of globalization'. The chapter tracks the trajectory of Cambodia's industrialization and insertion into the global economy over three interrelated phases: first, the beginnings of export-orientated garment production in the mid- to late 1990s; second, the promotion of Cambodia as an 'ethical producer' from 1999; and third, a more recent period in which he argues that manufacturers and local governments have privileged competitiveness in global value chains over labour compliance. In doing so, the author analyses the complex intertwining of global production, the genesis of the unique ensemble of actors in Cambodia, including the ILO Better Factories Cambodia programme and the anomaly of Cambodia's labour movement.

Also focusing on Better Factories Cambodia, Brown, Dehejia and Robertson analyse compliance data collected by the programme in Chapter 10, focusing on the factors that drive firms' decisions to become and remain compliant with international and national labour standards, arguing that it is not only the external pressure exerted by Better Factories Cambodia and by international buyers that stimulate compliance decisions, but also likely endogenous factors related to a business case for labour standards.

Finally, in Chapter 11, Bair and Gereffi ask what Better Work can offer producers in a country like Nicaragua and how the relatively high level of institutional development and workplace protections in that country affect the willingness of manufacturers to cooperate with Better Work Nicaragua. Drawing on data collected by the authors during fieldwork in the country, as well as on secondary literature, this chapter explores the contemporary context of the Nicaraguan industry, outlining the opportunities and challenges confronting the Nicaraguan apparel sector. Nicaragua is the first and only Central American country to participate in Better Work. The second poorest country in the hemisphere, Nicaragua's manufactured exports are heavily concentrated in apparel. While Latin America's share of the US apparel import market declined in recent years, Nicaragua has proven to be an exception to this rule, as export growth has remained robust. However, Nicaragua's strong performance relative to other regional exporters reflects the special benefits it received under the Dominican Republic–Central America Free Trade Agreement (CAFTA-DR) with the United States. Although these preferences have enhanced Nicaragua's competitiveness vis-à-vis other regional exporters, they are set to expire in 2014. As a result, because Better Work Nicaragua is being implemented during a period of uncertainty for the industry, it is critical to understand what local stakeholders in both the public and private sectors believe will be the consequences of this change in the regulatory regime, and how they are trying to respond to it. The chapter concludes that there is a crucial need for crafting a programme that reflects the specific conditions that characterize Nicaragua and differentiate it from other Better Work countries.

We then reflect on the key messages of the edited volume in the Conclusions and highlight directions for future research as well as identify policy recommendations.

Part I

1

Re-embedding the Market: Global Apparel Value Chains, Governance and Decent Work[1]

Frederick Mayer and John Pickles

Introduction

Much of the global economy is now organized in global value chains that span continents and connect producers and buyers across countries. This complex and dynamic structure of economic activity has rapidly spurred the productive capacity of developing and emerging economies. Economic globalization has created a great deal of wealth for some people in some regions of the world, but rapid economic growth has also been accompanied by the expansion of precarious forms of work, particularly at the margins of the new global economy, but also within core functions and companies. It has also led to pressure on labour in more traditional centres of production.

Our concern in this chapter is with the role of governance institutions in promoting decent work in apparel global value chains. By 'governance' we refer to those institutions that constrain or enable market actor behaviour – both in the public sphere, in the form of governmental policies, rules and regulations, and in the private sphere, in the form of social norms, codes of conduct adopted by businesses, consumer demand for social responsibility or other non-governmental institutions and social movements. We develop a very specific argument about the relationship between markets and governance. This argument builds on the work of Gereffi and Mayer, who theorized that globalization created a global 'governance deficit', to which states and societies responded by seeking to fill the gap with new governance capacities (Gereffi and Mayer 2006). Our analysis, like theirs, draws on Karl Polanyi's concept of a 'double movement'. In our reading, instead of focusing on the disembedding of markets from social control, we turn to the ways in which states and society are always engaged in struggles to re-embed market actors – in part – through new or renewed institutions of governance (Polanyi (1944) 2001). We see this process as a dynamic one of constant adjustment in which complex arrays of market actors respond to opportunities presented by the institutional environment, social actors respond by constructing new institutions to govern them, markets respond to the new institutional environments and so forth. There is nothing inevitable about the outcome of this on-going co-evolution, and the successful re-embedding of market institutions in forms of social regulation is not guaranteed, although it may happen both in the global North or South. Nor is it necessarily the case that this process of re-embedding will lead to coherent and integrated forms of state-led industrial policy, although in some parts of the global South state-led industrial policy seems to be increasingly on the agenda. Whatever the form such re-embedding takes, an important question concerns whether decent work and workers' interests will be adequately promoted and protected in the emerging architectures in both the global North and South.

Apparel is particularly interesting as a case study; it is highly globalized and a major global employer. Apparel is also a sector in which there have been considerable challenges of governance failure and widespread recognition of predatory sourcing and production practices, whether in the 1911 Triangle Shirtwaist fire in New York City or in the 2012 Tazreen Fashions fire in Dhaka, Bangladesh (Pickles 2012a). Our purpose in this chapter is to explore the extent to which new and diverse governance structures are emerging in global value chains that may, in some measure, be contributing to possibilities for creating improved working conditions in apparel production. The rapid rise of highly flexible global production networks, often involving informal or non-standard labour arrangements, has left many workers outside the control of effective governance institutions to protect their interests. In response, some workers and their advocates have generated sufficient pressure that increasing numbers of lead firms and industry forums have responded by experimenting with a variety of private governance initiatives. These experiments are particularly important, but their reach remains uncertain because they are occurring at the very time that companies, non-governmental organizations (NGOs) and worker movements have also been questioning the limits of private governance approaches (such as corporate social responsibility – CSR – and in-house or third party monitoring). As a consequence, the role for public governance in emerging industrial countries may, again, be becoming more significant (see Palpacuer 2006, 2008).

In the conclusion we return to the idea of the double movement. We suggest ways in which this perspective opens, rather than closes, options for thinking about ethical sourcing initiatives and improvements in working conditions and labour rights.

Globalization and the governance deficit

Gereffi and Mayer (2006) argued that governance institutions vary along three dimensions. The first is the distinction between *public* and *private* governance. Public governance is imposed by the state. Equally important, however, are the private, non-state institutions, ranging from broad societal norms to collective bargaining agreements that also establish the context in which market actors operate.

The second relates to the function of governance – whether to facilitate, to regulate or to redistribute (or compensate). *Facilitative governance* enables markets to form, creates jobs, attracts investment and otherwise promotes economic activity. *Regulatory governance* constrains the behaviour of profit-seeking firms that might otherwise tend to exploit workers, leading to poor working conditions, lack of job security, constraints on worker organization and general downgrading of industrial relations systems and practices. *Redistributive (or compensatory) governance* seeks either to redress inequalities through progressive taxation and social services or to compel firms to offer more generous wages and benefits.

The third dimension is the scale of governance – whether local, national or international, whether firm- or industry-specific or more cross-cutting. Public governance is more developed at the level of the national state and is typically non-sectoral. Private governance, on the other hand, may transcend national boundaries and be organized along sector- or even firm-specific lines.

In capitalist countries, institutions of governance – facilitative, regulatory and redistributive – initially arose under the aegis of the state as defensive measures responding to pressure from organized labour against predatory forms of industrialization. The modern welfare states that later arose out of these struggles shared a common basic commitment to market capitalism coupled with labour and other regulation and various forms of social protection, as well as reasonably comparable wage rates and standards of living. Despite the absence of strong international regulatory and distributive institutions, common national level governance institutions allowed for the temporary stabilization of international commerce among developed countries, a system Ruggie called 'embedded liberalism' (Ruggie 1982).

Globalization destabilized this regime and its 'virtuous circle' of regulated exploitation and 'shared' surplus. In this post-Second World War regime the share of profits going to wages increased. With subsequent globalization of production, the share of profits going to wages has declined year over year. The liberalization of international trade, the de-regulation of emerging market economies and the opening of former closed economies in Central and Eastern Europe, China and South-East Asia rapidly expanded the global market to incorporate countries with not only lower wage rates and standards of living, but also very limited governance capacities. The expansion of production networks beyond national borders, with attendant integration of sourcing and supply chains, posed challenges for those interested in sustaining and expanding the hard-won regulatory protections that guaranteed some level of decent work and protected workers against predatory practices in global markets (Pickles and Smith 2010).

The spread of production to lower income economies meant that an increasing proportion of work for the global market took place in locations where governance capacities were weak, if developed at all. Indeed, to attract international investment, newly industrializing countries sought to make themselves more attractive by strengthening investment protections and other policies intended to promote industry, while at the same time deregulating and privatizing production. Special economic zones in emerging reform economies, such as Cambodia, China, Thailand and Viet Nam (as in their Latin American *maquiladora* equivalents in the 1980s and 1990s), represent an extreme form of this strategy (Arnold and Pickles 2011). Often situated in export-processing zones and dependent on migrant labour, they maintain low-cost, disciplined and often unregulated and illegal workers in a regime of what Pongsawat (2007), describing border production in Thailand, has called 'partial border citizenship'.

At the global level, the strongest international institutions focused on market facilitation (most notably the International Monetary Fund (IMF) and the World Trade Organization (WTO), but also to a large extent the World Bank and the World Intellectual Property Organization). From the 1970s, the Bretton Woods institutions rapidly expanded their scope and reach, opening market access while resisting claims from trade unions and other non-state actors for embedded labour protections. Those institutions whose mission was labour protection – such as the ILO – were notably disadvantaged from the 1970s onwards, with the shift from tripartite forms of national industrial relations agreements to increasingly globalized, neoliberal policies of deregulation and market-driven mechanisms.

Gereffi and Mayer argue that this round of economic globalization created a global 'governance deficit', characterized by limited capacities in the emerging economies, weak international institutions, increasingly challenged institutions in advanced industrial countries and everywhere greater emphasis on facilitation than on regulation or redistribution. This deficit can be understood, in Polanyian terms, as a relative dis-embedding of the market from institutions of governance and enhanced opportunities for predactory behaviour.

The second movement: social responses to globalization and governance innovations

The story of the further disembedding of hard-won governance mechanisms with the expansion of global value chains has been well documented. The result has often been struggles over the appropriation and distribution of social surplus, what Gibson-Graham et al. (2000) call 'class processes', which in turn generate Polanyian-like countermovements to re-embed the economy in practices and institutions of social regulation, as governments, NGOs, organized labour groups and other social actors strive to create new institutional arrangements to re-regulate product and workplace standards and worker rights in the global economy, what Vorley et al. (2007) have called 'regoverning markets' and Webster et al. (2008) referred to as 'grounding globalization'.[2] Thus, for example, by the late 1990s, the dominance of the Bretton Woods institutions (and the Washington Consensus they seemed to represent in re-shaping global governance[3]) had generated strong counter-currents among anti-globalization, alter-globalization and global justice movements. Together with strong critiques of neoliberalism, these resulted in sustained efforts to change the ways in which the inter-governmental agencies dealt with labour and human rights issues. In 1999, oppositional movements poured onto the streets of Seattle in what became an iconic example of the struggle over global governance (repeated several times since, including at G8 meetings in Genoa, WTO meetings in Geneva, World Bank and IMF meetings in Washington, DC, and Summit of the Americas meetings in Miami and Quebec City). As Pigman (2005: 1) suggested:

> the emergence of widespread, organized resistance to the discourse of neoliberal globalization at the end of the 20th and beginning of the 21st centuries indicates how the fruits of global trade itself have empowered effective strategies of resistance to the widening

> disparities of wealth and power that global markets have perpetuated. Global social movements have challenged the various institutions maintaining the parameters of the neoliberal global trade regime – the WTO, the IMF and World Bank, the OECD, the World Economic Forum – to 'civilize' world trade by re-embedding it within a sustainable global framework of human social relations.

As a consequence, it is crucial in analysing global value chains to assess the strategic importance of the social pressures unleashed by globalization and the proliferation of institutions and actors reacting to them. In some cases, these are currently driving innovations in governance – both changes in existing institutions and new forms of governance. In this way, states and international organizations, labour unions and firms, NGOs and citizens are responding to the dis-embedding of economic practices from social systems, and each seeks in one way or another to create some basic level of social or brand protection in the global economy. These innovations are both public and private, operate on international, national and local levels, and seek either to spur economic or social upgrading, or both. They also have very different levels of scope and effect.

Private governance responses

Apparel is a particularly interesting sector because of the diversity of positions taken by actors in its global value chains (Appelbaum 2005; Keohane 2002; Tickell and Peck 2003). Companies such as Levi's, Marks & Spencer and the Cooperative have long operated on deeply embedded commitments to ethical sourcing, high quality and localized production. By contrast, mass merchandizers such as Walmart came to typify the footloose and delocalized predatory outsourcing of the new global economy (Appelbaum 2004; Appelbaum et al. 2005; Gereffi and Christian 2009; Oka 2012). Yet other companies (such as Nike and Gap) that were once the poster-firms for such footloose sourcing and the sweating of labour that followed, have now emerged as industry leaders in transparency, monitoring and CSR. The pressure on visible brands to ensure supply chain transparency and compliance has increased dramatically in recent years. For Nadvi and Wältring (2001: 34), despite their toothless nature, core labour standards have become a model for private social standards. More recently, Nadvi (2008: 323) has suggested that 'Compliance with international standards is now a *sine qua non* for entry into globalized production networks' (see also Committee on

Monitoring International Labour Standards 2004 and Robertson et al. 2011). The change has been significant and is potentially far-reaching, with a company such as Walmart now aggressively positioning itself as a socially and environmentally responsible company.

Particularly with the deregulation of public forms of governance and the broader governance deficit, social pressures generated by globalization have found expression in a variety of private governance experiments. Efforts outside of state agencies to regulate firm behaviour and compliance in their value chain have ranged from largely voluntary efforts to expand CSR across the supply chain, to activist and consumer pressure for ethical sourcing and 'clean' clothes, to investor movements to create markets for social responsibility. These are all part of a broader movement of what Appelbaum (2005: 373) has called 'the privatization of labour enforcement'.

Private governance facilitates markets through standards adopted by industry groups, the activities of business promotion groups, such as chambers of commerce, and framework agreements that establish norms of trust and conduct. In the context of global value chains, this might include efforts by development associations to attract certain forms of investment or to cooperate with greenfield startups (see Bair and Gereffi, Chapter 11 in this volume). It might also include direct changes in production process, in the structure of buyer-driven value chains or CSR initiatives by leading brands as part of broader demand management strategies.

Private governance can have a significant regulatory impact in the form of policies adopted by firms beyond any minimum standards required by law, negotiated arrangements between firms and labour advocates, and pressure campaigns driven by civil society and consumers. It can also influence distribution through voluntary or semi-voluntary agreements by firms to pay living wages, through negotiated wage agreements, through voluntary provision by firms of other benefits and by social institutions that provide services to the unemployed and working poor.

Codes of conduct, supply chain monitoring and social labelling

One response to the globalization of sourcing networks and the intensification of pressure from consumer groups has been the (initially begrudging) acceptance on the part of global buyers of their own responsibility to ensure that working conditions meet basic standards. These businesses developed a wide range of codes and monitoring systems, many involving a certification programme of some kind. In the United States, the most widely used certification programmes are SA8000 (a social accountability standard that, inter alia, requires a 'basic needs wage' and associational rights), Worldwide Responsible Apparel Production (WRAP), which operates a factory certification programme, and the Fair Labor Association (FLA) which grades members' suppliers on their ability to meet code requirements.

Certification has worked best where image in the market place is central to the success of large brands, as exemplified by the case of Nike (Mayer and Gereffi 2010). Since its rapid rise to become one of the world's leading brands, Nike has defined what it means to live on an image. Its size, visibility and importance as a consumer image meant that Nike became a flashpoint for consumer campaigns, particularly of the anti-sweatshop movement. In response, Nike drafted the Nike Code of Conduct, aimed at improving working conditions in its contract factories.

Perhaps the strongest certification regime operates in a particular niche within the apparel industry, that of US Collegiate Apparel (Bair 2008). Since the 1990s, there have been intense discussions among student organizations, universities, monitoring agencies and companies supplying collegiate apparel about the need to monitor supply chains and conditions of work. Out of the anti-sweatshop student movement of the 1990s, in addition to the FLA, the Workers Rights Consortium (WRC) was formed to respond to the same concerns and pressures of colleges and universities. Companies that sign on with the WRC must agree to provide access to their suppliers and permit unannounced and independent monitors to carry out spot checks at their sourcing facilities in direct response to workers' or third party complaints.

In recent years, both industry and anti-sweatshop organizations have begun to question the viability of monitoring and voluntary code enforcement as an effective model of protecting workers and guaranteeing safe and fair working conditions, even in the production of goods for relatively high-priced markets, such as collegiate apparel (Barrientos and Smith 2007; Jenkins 2001; Mamic 2003). Indeed, to listen to many buyers and producers, it is not the lack of governance mechanisms that concerns them, but the *surfeit* of regulatory demands and codes with which they and their suppliers have to deal. Buyers and suppliers complain about 'monitoring fatigue' or 'code overload' – where factories may be subject to as many as 30–40 annual inspections and may be required to meet the different code requirements of national governments and the eight to 12 brands for which they produce.

Frustrations with the first generation of codes and monitoring programmes to protect workers, guarantee sweat-free working conditions and maintain clean supply chains have led to a variety of ethical sourcing, sustainable compliance and social labelling initiatives (Hughes 2006; Roberts 2003). Branded textile and clothing firms have emerged as niche suppliers of ethically sourced fibres, fabrics and clothing (Hughes et al. 2008). Aimed at driving ethical sourcing, transparency and compliance down through the entire supply chain, ethical sourcing initiatives are usually also linked to consolidation of the supply chain and reduction in the number of suppliers. Together these allow much greater private control of suppliers and public transparency and oversight along supply chains. In the main, these ethical sourcing initiatives are driven by buyer concerns for brand reputation and market access. Thus, for example, in response to continued pressure from student and other NGO groups and in the face of continued problems with supplier compliance, Nike disclosed its entire list of suppliers in 2005. In May 2007 it released for the first time its auditing tools and announced, along with its FY 2005–6 corporate responsibility report, a series of new initiatives in CSR, and expansion of its soccer, community and environmental recycling programmes.

Collegiate apparel supply chains are again instructive. Here the situation came to a head in 2005 with the ending of MFA-quotas and threats of lost contracts and closure to precisely those factories that had responded to earlier pressure to raise standards, improve working conditions and – in some cases – increase wages. The ensuing deliberations among the WRC, United Students Against Sweatshops, the FLA, universities and their licensees have increasingly revolved around proposals for an alternative sourcing model, called the Designated Suppliers Program (DSP), and a new monitoring programme, called FLA 3.0.

The DSP is an attempt to move from monitoring to partnerships based on an ethical sourcing model and positive incentives for compliance with labour standards. This model is intended to replace the 'race to the bottom' with a 'race to the top', a principled restructuring of licensed apparel supply chains in which strategic suppliers who maintain standards and meet basic criteria will be rewarded with longer term contracts, technical assistance and higher prices (WRC 2010). The programme has gained support from some universities, but the failure of the Bush administration Department of Justice to supply an antitrust letter delayed broader implementation and may have effectively blocked the initiative.[4] In the interim the WRC has been working on the living wage model in conjunction with Knights Apparel and its Alta

Gracia factory in the Dominican Republic. In partial response to growing pressure to adopt the DSP, the FLA's new approach to monitoring, FLA 3.0, aims to create a system of 'sustainable compliance', in which participating factories are given assistance to monitor and improve their own performance, while simultaneously disclosing the suppliers in their collegiate supply chain.

For other companies, the turn to ethically sourced products is driven less by the need to manage reputational risk and is focused more on social labelling approaches. Whereas supply chain monitoring is more of a defensive measure designed to protect the brand from association with negative images, but usually not something that is advertised directly to consumers, social labelling seeks to make socially responsible production integral to the brand, part of what a consumer is buying. For example, Fair Trade clothing, launched in 2010 by Transfair, certifies clothing as 'fair trade' if it meets certain standards. A second example that combines both brand identification and social labelling is PeopleTree, which in 2009 launched a new line of clothing 'bringing fair trade and organic fashion to the 16–24 age group' (Holmes 2009). A third example is School House clothes, started by a former Duke University student, which partnered with ALARM, a Sri Lankan coalition of labour rights organizations and trade unions, to launch a line of ethical fashion aimed at university students.

Such programmes are not without their critics. Writing recently about India, Seidman (2009) has offered a critique of such social labelling programmes by suggesting the need to consider ways in which social rights and logics of citizenship might be extended:

> Social labelling programs are organized around the logic of the market, appealing to global consumers by certifying that producers have cleaned up their supply chains. But if India's goal is not simply an export-promotion strategy, perhaps they should look to programs based instead on the logic of citizenship, expanding schools and programs to support poor families rather than offering exporters a certificate that turns compliance with child labour laws into a marketing strategy.

Gap's community development project in Mewat, India, is aimed at precisely this kind of broader community involvement, service and education provision, and improved working conditions in local communities. It is a collaborative project between Gap India, NGOs, community groups and female garment workers, and is aimed at trying to find a new way to deal with problems of child labour in the supply chain. Cognizant of past failures of monitoring and sanctions to deal with child labour in subcontracting, Gap is currently experimenting with precisely the kind of educational and infrastructural programmes Seidman has called for (Gap 2009). But, as Seidman suggests, such one-off private initiatives are inadequate to the task of managing complex global supply chains in an ethical manner, and despite the benefits they have for specific workers and factories individual initiatives do not resolve the challenge suppliers and workers must deal with in the face of multiple codes, standards and audits from their various buyers.

The Global Social Compliance Programme (GSCP) is one possible private governance response. GSCP is a private sector initiative that seeks to create a common platform of standards and codes that member brands agree to implement throughout their supply chain, ' harmonizing existing efforts to deliver a *shared, global and sustainable approach* based on consensus and best existing practice'(see http://www.gscpnet.com/).

International framework agreements

Workers, of course, have always responded in diverse ways to the circumstances in which they find themselves. The nature of the capitalist employment contract is inherently one in which the organization of production, the nature of work and level of remuneration are contested. How we understand the positional power of labour to shape its own conditions of work and contribute to the creation of decent work depends in large part on the specific form of political-economic regime, the industrial sector, the structure of the value chain and the ways in which a particular industry and regional economy are inserted into global production networks.

International framework agreements (IFAs), negotiated by the global union federations with multinational enterprises, constitute an important example of worker response to globalization. These agreements typically seek to hold multinational enterprises to ILO core labour standards throughout their production networks. For example, in 2007, Inditex (the owners of Zara) and the International Textile, Garments and Leather Workers Federation (ITGLWF) signed an IFA to reinforce worker rights to form and participate in free trade unions, to bargain collectively, and to ensure that Inditex's code of conduct improves workers' conditions within Inditex's supply chain (Miller 2011). Inditex agreed to inform all of their suppliers of this agreement, and the ITGLWF agreed to notify its member unions worldwide. Together they develop worker education programmes focusing on rights of association and collective bargaining. At the time, the NGO, Impactt, commented: 'To our knowledge this is the first agreement of its kind in the garment industry and it represents a major step forward for Inditex and ITGLWF and the whole area of labour rights' (ITGLWF 2006; Buttle 2007). Miller et al. (2011) describe agreements of this kind as essential elements of mature systems of industrial relations, as they respond to inherent weaknesses in social auditing and codes of conduct by recognizing the 'enabling rights' of collective bargaining. IFAs have been increasing in number and scope in other sectors, but the IFA between Inditex and the ITGLWF remains the only one in the apparel industry. Moreover, even in this case the enforcement of the IFA largely depends on ITGLWF (now IndustriALL) associates at the country level and whether they are actually present in factories. Where there is no presence of ITGLWF-affiliated unions in apparel factories the IFA has not been enforceable.[5]

Limits to private governance

Notwithstanding the diversity of private governance initiatives, there are clearly limits to such approaches as stand-alone alternatives to public governance. The impact of private governance mechanisms on general working conditions and wage stability – crucial criteria for success in responding to the regulatory vacuum for global production networks – is far from demonstrated, although their effects in specific chains and factories may be significant. Private monitoring and enforcement programmes, in particular, have received a number of criticisms in recent years. These include:

- companies are monitoring their own suppliers (see O'Rourke 2011);
- monitoring is expensive and the costs are often borne by the supplier;
- supplier factories must comply with codes and monitoring audits for every brand they produce for;
- even among factories attempting to be compliant, factory violations are produced by last-minute changes in fashion, design, fabric and/or colour or unrealistic pricing and production deadline requirements imposed by buyers (see Plank, Rossi and Staritz, Chapter 5 in this volume);[6]
- compliance efforts are threatened by conditions of global recession, currency volatility and financial and order uncertainty;
- monitoring seems ineffective in changing the practices of non-compliant producers; and
- suppliers tend to comply only with more easily adopted 'low hanging fruit', rather than deeper structural issues and enabling rights.[7]

Private governance has been shown to be most effective when there is a clear lead firm with considerable power in its value chain, when collective action by consumers or advocacy groups is possible, and when social upgrading aligns well with commercial concerns (Mayer and Gereffi 2010). It seems to be particularly enhanced when firms foster regular opportunities for worker representation and participation in decisions about work schedules, practices and responsibilities (Locke et al. 2007). IFAs are promising, but still very limited and depend fundamentally on strong enabling environments for labour unions. Ethical sourcing is still in its infancy and remains something of a niche phenomenon, although it may have important demonstration effects that destabilize industry claims about what is economically feasible in contemporary global supply chains. Consumer demand for products that are produced in compliance with enhanced labour standards is also relatively limited, but has been important in particular markets, such as collegiate apparel.

Public governance responses

After more than 30 years of state and international deregulation, and in the face of increasing evidence of a race to the bottom in global supply chains, the question of public governance has again become more important. It is particularly interesting to see the proliferation of public governance responses to nomadic sourcing and footloose predatory production practices. These initiatives are taking place at both the international level, in global or regional institutions, and at national and local levels.

Global and regional responses

At the global level, the major institutions governing the international economy, notably the Bretton Woods institutions, were initially reticent to respond directly to concerns about social downgrading and predatory working conditions. As noted earlier, global public governance has historically been better at facilitation than at regulation or redistribution, although the ILO has always been an important voice for workers and for promotion of decent working conditions.

One recent significant development is the shift in World Bank policy, which now requires adherence to ILO core labour standards in projects it funds, a standard applied by the World Bank's private sector lending arm, the International Finance Corporation (IFC), since May 2006 (Ebert and Posthuma 2011). What real effects of this commitment and what 'respect for core labour standards' will mean at the level of implementation is still to be determined. Indeed, the Bank's initial formulation of the Doing Business benchmarking programme considered labour regulations, such as hours of work, minimum wages, advance notice of mass dismissals and protection against discriminatory practices, to be undue impediments to 'doing business', resulting in higher scores for countries that made their labour legislation more flexible. These indicators have been subsequently revised by the World Bank, but their initial inclusion called the strength and sincerity of the Bank's commitment to core labour standards into question, particularly among organized labour groups.

The effort to link labour provisions to multilateral trade agreements at the WTO has similarly been limited in scope and effect. Although there was a push in this direction in the Doha Round negotiation, advocates initially for including some type of labour provisions in WTO negotiations met strong resistance. This came largely from developing countries, who viewed such efforts with suspicion as disguised protectionism that could act as potential barriers to their own further development (WTO 2010a).

What cannot be accomplished at the global level may be possible at the regional level. Many regional institutions of governance play important roles. In Europe, regional public governance, in the form of the European Union (EU), is considerably stronger than global institutions, most notably in their regulatory and redistributive roles. Regional development banks, such as the Inter-American Development Bank (IADB) and Asian Development Bank (ADB), are often more attuned to the social dimensions of globalization than the World Bank. Linking labour provisions to regional FTAs is also increasing. Given the proliferation of FTAs, such linked labour provisions could play a useful role in the absence of a broader multilateral approach.[8] The EU has done most in this regard, with regional trade agreements aimed at fostering economic development and political openness and bilateral trade concessions such as 'GSP+' explicitly granted to beneficiary countries applying ILO core labour standards and other basic rights. In North America, regional free trade agreements such as the North American Free Trade Agreement (NAFTA) and the Central America Free Trade Agreement (CAFTA) do include side agreements on labour, but their provisions are very weak and they do not include explicit references to ILO labour standards (Mayer 1998). The United States (particularly during the Bush administration) has placed even less emphasis on social concerns in its bilateral trade agreements.

State responses

Notwithstanding the innovations that have taken place in linking labour provisions to trade agreements, integrating labour standards into the policies of the Bretton Woods institutions, and building new international institutions from the South, significant obstacles remain in the development of effective inter-governmental regulatory or distributive governance capacity at the global or regional level. In the public realm, governance capacity remains heavily concentrated at the level of the national state, but state-level governance has been severely eroded by neoliberal commitments to deregulation and market-based mechanisms, particularly as production moves to low-cost producing locations in the global South with less well developed facilitative, regulatory and compensatory capacities.

States have, however, begun to respond. Although much of their effort has been focused on industrial policy and economic upgrading, many emerging economies have begun to adapt their existing state institutions and develop new governance capacities to regulate production processes and, to a lesser extent, to provide social safety nets and other forms of compensatory mechanisms. In recent years, governments have come to power in Latin America, Africa and elsewhere that resist Bretton Woods prescriptions for economic reform structural adjustment, and the debt burden it often entails. In Latin America, governments in Argentina, the Plurinational State of Bolivia, Brazil, Ecuador, Nicaragua and the Bolivarian Republic of Venezuela seek to create the possibility for autonomous alternatives to global neoliberal models of free trade and open markets, each mobilizing state resources and national institutions to foster economic and social development. Thus, Piore and Shrank (2006: 2) report that:

> Brazil, Chile, Costa Rica and the Dominican Republic have rededicated themselves to labour law enforcement in recent years. And potentially more fundamental reforms are underway from Argentina, where they are motivated by domestic party politics, to Central America, where they are a product of transnational pressure emanating from the campaign for a US-Central America FTA.

In Bangladesh, deep social unrest driven by low wages and poor working conditions led in 2010 to a coalition of major apparel buyers joining labour advocates in pressuring the government for a major increase in the minimum wage for apparel workers. In response, Bangladesh increased its minimum wage in the apparel sector by 80 per cent

(Pickles 2012b). The increase was, however, far below what had been demanded by workers, and weak trade unions continue to hamper the ability of workers to press for further gains (Anas 2011).

Cases of institutional reforms in post-socialist states may also be instructive. In these countries, the roles played by former state socialist institutions have changed quickly in recent years. Planned economies were deregulated quickly after 1989, state institutions were weakened and manufacturing industries suffered massive retrenchment. With the resulting loss of long-established markets and trade links, export-oriented industries declined, employment was shed rapidly and plant and capital were redirected into private hands, sold or abandoned.[9] But most post-socialist states retained the working (albeit weakened) institutions of state socialism (labour inspectorates, health inspectorates, working hours law, overtime regulations, insurance and pensions requirements, etc.). By the late 1990s and early 2000s some governments (such as Bulgaria and Slovakia) were providing them with additional support and funding to ensure that basic working conditions were better regulated. While many cases of workplace abuse and hyper-exploitation have been reported (most notably in Bulgaria by Clean Clothes Campaign), in general the legacies of labour inspectorates, health and safety inspectorates, respect for working hours, child labour laws and wage payments and contract conditions, remained important strategic institutions for workers and partial regulators of predatory business practices (Pickles et al. 2006; Smith et al. 2008; Chapter 5 in this volume).[10]

Precisely how such state institutions are able to function or be revitalized as effective bodies of workplace regulation requires much more investigation. The paradox of state regulatory regimes exercising their jurisdiction differentially across firms in the same industry points to the importance of understanding how such regimes are articulated with international norms and private firm codes of behaviour and practices. It requires a much fuller understanding of the roles played by workers' firm- and regional-level struggles for better working conditions in shaping government policy. China is an important example in this regard (see Box 1.1).

Box 1.1 China's Labour Contract Law

In response to the increasing deficits carried by state-owned enterprises, social unrest among the rapidly growing number of factory workers in export processing zones and international pressure from NGOs, trade unions and some major buyers concerned about brand reputation, China initiated far-reaching reform in its labour laws. These were already signalled in 2005 with the state-sanctioned naming of the Chinese Year of Corporate Social Responsibility, a commitment to harmonious work and society that has been consolidated and expanded in subsequent years (see China CSR 2010). In January 2008 the new Chinese Labour Contract Law came into force (see Lan and Pickles 2011). This was developed in response to growing public concern about the mistreatment of contract workers, the growing numbers of cases where managers failed to pay proper wages and the rapid increase in seasonal and temporary (often migrant) workers attendant on the demise of state-owned enterprises, the former bastions of stable regional employment in many towns. The law aimed to re-assert the basic values of labour protection in state policy, and specifically to deal with some of the ways enterprises were getting around the existing labour laws. With increasing financial pressure on state-owned enterprises and weak regulations for private firms, employers were increasingly avoiding payment of benefits. They employed workers on probationary contracts, or they avoided offering permanent contracts by employing workers on a series of short-term repeating contracts. Under the new law, employees were given formal term contracts; they can only be terminated with cause; and after a maximum of two term contracts the employee must be given an open-ended permanent contract, which ends only with the employee terminating his or her contract, termination for just cause or retirement. Under the new law, individual workers and the worker's union are also able to go to court to enforce their rights. One provision of the new labour law requires negotiations between workers and companies over the terms and conditions of work and enhances the role of the All China Federation of Trade Unions (ACFTU) in China's labour relations system.

Over the longer term, the law may have important consequences, for workers in China and elsewhere (Global Labour Strategies 2008). The law has encouraged workers to fight for their rights, with a rapid increase in the number and strength of labour disputes. In 2007, China's labour dispute arbitration committees accepted 350,000 cases, an increase of 10.3 per cent from 2006 (CLB 2009: 14). The same survey revealed that the average number of workers in disputes had risen to 28.6 in publicly-owned enterprises and 51.3 in privately-owned enterprises. During the first quarter of 2008 alone, the labour courts in Dongguan, Shenzhen and Guangzhou (the three industrial cities in the Pearl River Delta) accepted more than 10,000 cases, double the number over the same period the year before (Wang et al. 2009: 492). As labour costs have risen and manufacturers have responded with layoffs, increased labour contracting, 'tricking the contract' and – in more extreme cases – relocating factories to provinces or countries with lower costs (Zhu and Pickles 2013). The result is the stabilization of working conditions for core working and the increased level of precarity for millions of low-wage temporary contract workers throughout the country, particularly with the rise of non-renewable three-month contracts. The Labour Contract Law also changes the ways in which multinational companies must deal with worker organizations, leading to sourcing shifts as buyers manage risk by complementing orders from China with orders from other low-cost regional producers, such as Cambodia Indonesia and Viet Nam.

Public-private governance responses

Although it is useful to distinguish between public and private governance, it is also the case that the effectiveness of each, in the end, depends on the nature of the other. Private governance, in the form of collective bargaining agreements, for example, inevitably depends on the extent to which labour law protects workers' rights to organize. Similarly, the effectiveness of state-promulgated regulations depends in large measure on the willingness of firms, unions, NGOs and consumers to observe them. For the same reason, some of the more intriguing governance innovations are explicitly hybrid, with elements of both public and private. The ILO Better Factories Cambodia project, now expanded to the Better Work programme of the ILO and the IFC, is one such example.

The Better Work programme

The Better Work programme, a joint initiative of the ILO and the IFC, stems out of the Better Factories Cambodia (BFC) project. In this initiative, the Cambodian export apparel industry has been reorganized around a combination of private self-regulation with national and international public interventions (Polaski 2006). This collaboration originated with the 1999 US–Cambodia Textile Agreement, which created an incentive for the Cambodian government to improve labour conditions in its factories in order to obtain greater access to the US market.

In this initiative, the ILO 'combines independent monitoring with finding solutions, through suggestions to management, training, advice and information' (ILO 2010). Buyers appear to have found the programme valuable as a form of 'reputation risk insurance', which has proven critical to the ongoing success of the experiment. The Cambodian government has had to intervene to prevent free riding because the ILO monitoring programme is factory-based, while the quota bonus was awarded to the country as a whole, based on overall performance. Nevertheless, the Cambodian government and the country's apparel manufacturers agreed to continue the strategy indefinitely, even after the quota system expired at the end of 2004. The ILO continues to supervise the monitoring project as manufacturers continue to maintain orders through the branding and imaging of the Cambodian apparel industry built up around these preferential trade incentives. However, with China's new policies on CSR and labour contracts and workplace rights, Sri Lanka's initiatives on ethical sourcing, and other similar efforts, Cambodia is no longer a unique case of a state commitment to ethical production. As a result, the focus in Cambodia has expanded to increasing competitiveness by reducing factor costs through infrastructural development, improving business climate by easing import/export fees and reducing the steps in the process, eliminating corruption, increasing workers' skills and other issues that were not adequately addressed during the quota period (but see Arnold in this volume). Such efforts are driven largely by the fact that between 2005 and 2009 roughly 75 per cent of Cambodia's exports to the United States (its primary market) were concentrated in China safeguard categories (mostly cotton knit shirts and trousers), which presents serious risks to the industry as safeguard removal takes effect (see Arnold and Toh 2010). The ILO itself has noted the increasing use of short-term flexible labour contracts (particularly Fixed Duration Contracts) since 2005 (ILO 2006). The global financial crisis and recession after 2008 have had serious effects. After years of growth, apparel exports dropped sharply as demand from the major export markets dried up and buyers appeared to shift to cheaper suppliers in Viet Nam and Bangladesh (Better Factories Cambodia 2009), although Brown et al. (2011, 2012) have found some evidence that being compliant was a key factor helping some factories to survive the crisis since 2009.

Notwithstanding concerns about the sustainability of the Cambodia experiment, the programme has now evolved into a larger effort, the ILO/IFC Better Work programme, involving apparel production in Haiti, Indonesia, Jordan, Lesotho, Nicaragua and Viet Nam, with prospects for further expansion. As in Cambodia, the programme involves multiple stakeholders, including 'relevant employers' and workers' organizations, the national government and international buyers. The parties have agreed to allow Better Work to monitor compliance with international labour standards and to provide both transparent reports on progress towards meeting those standards and technical assistance.[11] As other chapters in this volume show, there is increasingly clear evidence suggesting that these efforts have had direct and positive effects on working conditions and business performance among the participating factories.

Conclusions

Our conclusions in this chapter are necessarily tentative, but our reading of the evidence suggests five points.

First, the 'double movement' has often been interpreted as an 'action-response' or a 'pendulum' model of societal response to disembedded market actors. An alternative reading of the double movement would be one that was more sensitive to the complex ways in which economic decisions are made among a variety of actors facing diverse opportunities and constraints. In this sense, the double movement would be understood as an ongoing dynamic, in which markets and governance institutions constantly adapt to changing circumstances (Gereffi and Mayer 2006; Evans 2008; Piore 2008; Palacios 2001; Turner 2007). It is also an ongoing (i.e. hourly and daily) struggle by workers, worker organizations and state regulators to fight for better working conditions and strive to increase (or at least maintain) the share of profits going to wages and benefits.

Second, although institutions of private governance are perhaps more well-developed for apparel than for almost any other sector, there are significant limits to the models that have developed to date. Codes of conduct and certification, in particular, arose at a particular moment in which market power was heavily concentrated among branded buyers sensitive to the concerns of consumers and NGOs largely located in developed countries. As the markets quickly expand in the global South and new or expanded forms of consumer demands and firm behaviour emerge, this form of private governance will need to adapt.

Third, although there is considerable policy logic to strengthening global institutions with the capacity to regulate or redistribute, the geopolitical reality is that progress on this front is likely to be very slow. The Bretton Woods institutions have responded to social pressures, but their missions and institutional cultures have considerable inertia. Stark differences among nation states and their continued sensitivity to sovereignty suggest dim prospects for breakthroughs at the global level. Regional agreements hold somewhat more promise, but deep and comprehensive trade agreements that embody labour protections are not likely in many regions of the world in the near term.

Fourth, the strengthening of state governance capacities, most notably in the emerging market economies of China, India and Brazil, as well as elsewhere, holds considerable promise. With the decline of neoliberal orthodoxy, governments are increasingly concluding that their desire for economic upgrading need not come at the expense of abandoning their regulatory and social protection functions. Indeed, some states are increasingly recognizing that workplace standards and enforcement mechanisms can be essential elements of value chain upgrading. How far these trends are likely to go, however, will depend in part on the extent to which governments are, in the end, responsive to domestic social pressures, and in part on their leverage vis-à-vis global market actors. Clearly, what might be possible for China because of its size might not be possible for Guatemala, for example, and so it will be necessary to develop scale-dependent metrics of effective governance institutions and practices.

Finally, re-embedding global markets will undoubtedly require hybrid and complementary institutions of governance, public and private, operating on multiple levels – global, national and local. Whether we are muddling towards structures that will enable reasonably stable and socially acceptable market structures is an open question, however. To answer it, we need to know much more about the temporal and spatial lags in the double movement; where and under what conditions the governance deficit leads to societal responses; and where and under what conditions labour and more general social pressures are translating into new and more or less effective governance.

Notes

1. Research for this chapter was supported by the Capturing the Gains Research Network. Preliminary versions of the paper have benefitted from review by and comments from Stephanie Barrientos, Gary Gereffi, Shane Godfrey,

William Milberg, Doug Miller, Dev Nathan, Anne Posthuma, Arianna Rossi and three anonymous reviewers.

2. Burawoy's (2010a) reaction to Webster et al.'s (2008) attempt to mobilize Polanyi in the context of firm-level and locally-based efforts to regulate the excesses of globalized production networks and the increasing insecurity of employment they have created has, in turn, generated a vigorous exchange among sociologists of labour on the uses of Polanyi in this context. See the recent issues of *Global Labor Journal* (Webster 2010; Lambert 2010; Caspersz 2010; Clawson 2010; Burawoy 2010b; Waterman 2010).
3. The term Washington Consensus refers to a series of reforms and prescriptions emerging in the 1980s that together became a 'standard' package for structural adjustment programmes in developing and late-socialist countries. Promoted by Washington-based institutions such as the IMF, World Bank and the US Government, the package included policies of macroeconomic stabilization, trade and investment liberalization, privatization, deregulation and an expanded role for the market in domestic economies. In Eastern Europe these structural adjustment measures were implemented with great speed on the grounds that only 'shock therapy' would create the deep-seated reforms necessary to sustain liberal market economies.
4. The US Justice Department has subsequently said it has no plans to pursue an anti-trust complaint, thus removing a major barrier to the broader acceptance of the DSP. See http://www.justice.gov/atr/public/press_releases/2011/278340.htm (accessed 11 February 2013).
5. We are grateful to Arianna Rossi for this insight. With the Rana Plaza collapse in Savar, near Dhaka, Bangladesh, on 24 April 2013 – the deadliest clothing factory accident in history with 1129 killed and many more injured – recognition of the breadth of non-compliance in Bangladeshi factories has led to several new public and private sector initiatives to regulate the industry.
6. Oxfam's (2004) report demonstrated clearly the ways in which efforts by human resource and compliance departments in large clothing companies to ensure code compliance in supplying factories were often undercut by costing departments that simultaneously squeezed down the price of contracts, demanded ever tighter production deadlines and quality standards, and adjusted orders and pattern-types during the production runs without any extension of deadlines or contract price. The result was an almost inevitable squeezing of work schedules and compliance goals.
7. See Barrientos and Smith (2007); and Posthuma and Nathan (2010).
8. According to the WTO: 'Regional Trade Agreements (RTAs) have become in recent years a very prominent feature of the Multilateral Trading System (MTS). The surge in RTAs has continued unabated since the early 1990s. Some 421 RTAs have been notified to the GATT/WTO up to December 2008 ... At that same date, 230 agreements were in force. If we take into account RTAs which are in force but have not been notified, those signed but not yet in force, those currently being negotiated, and those in the proposal stage, we arrive at a figure of close to 400 RTAs which are scheduled to be implemented by 2010. Of these RTAs, free trade agreements (FTAs) and partial scope agreements account for over 90%, while customs unions account for less than 10%' (WTO 2010b).

9. By the mid-1990s, the delocalization of European and US clothing manufacture had revitalized the industry throughout the region around stitch-up, CM and CMT contract manufacturing, structured largely by international buyers and manufacturers. As a result, the industry re-emerged quickly, but largely under the radar of the state authorities. These remained weak and the institutions of the state continued to be under-funded and under-incentivized (see Begg et al. 2003).
10. For documentation of abuses in Bulgarian supply chains, see Laleva and Musiolek's (1999) account of the 'Conditions in the Savina factory'. For discussion of the consequences of state socialist institutions on post-socialist working conditions, see Pickles and Smith (2010).
11. Better Work, available at http://betterwork.org/global/aspx (accessed 06 September 2013).

2

Economic and Social Upgrading of Developing Countries in the Global Apparel Sector: Insights from Using a Parsimonious Measurement Approach[1]

Thomas Bernhardt

Introduction

Historically, the apparel sector has played an important role as a pathway to industrialization and economic development. Today, as the production of goods and services is increasingly becoming fragmented and organized within international networks and global value chains (GVCs) – with lead firms coordinating suppliers, assembly, design, logistics, marketing and branding in multiple locations across the globe – economic development has more and more become associated with upgrading within such international networks. This is particularly true for the apparel sector, which is well-suited for such global production arrangements given that most intermediate products can be easily traded at each stage of the value chain. As a consequence, it has become one of the most globalized sectors in the world where production and trade are largely organized within and shaped by GVCs, which, in turn, are characterized by a high degree of involvement of developing country producers.

However, in recent years the global apparel sector has gone through two major crises (Gereffi and Frederick 2010). The first, at the end of 2004, was the expiration of the Multi-Fibre Arrangement (MFA), which had governed international garment trade since 1974 by imposing quotas on developing countries' apparel and textile exports to developed countries. Its end marked a deep structural break and led to substantial reshuffling in the global apparel value chain. Given its far-reaching implications, special attention will be given to this structural break in the analysis that follows. The second crisis is the global economic recession that has spread around the globe starting in 2008 – which further accelerated the reconfiguration and restructuring of the global apparel sector.[2]

The case study literature on the apparel sector that analyses these developments is extensive. These case studies provide a lot of in-depth information on the dynamics and on the (underlying drivers of) success or failure of apparel production in specific countries or regions. The purpose of this chapter is different. Using a parsimonious approach, it aims to provide an overview assessment of sector-wide dynamics in the global apparel industry during the last decade, and particularly after the MFA phase-out. Within this wider sectoral picture, this chapter will also discuss how a number of selected supplier countries have fared post-MFA and trace their economic and social upgrading and downgrading trajectories. This analysis will take a quantitative approach. It will be based on a narrow set of indicators and rely on secondary data. Such an approach, admittedly, has certain limitations but it allows us to make some inferences about overall trends and to compare performances across countries.

The methodology is presented in the below section. In the two following sections, this methodology is applied to identify economic and social upgraders and downgraders, respectively, among a sample of apparel-exporting developing countries. The 'overall upgrading' section investigates the relationship between economic performance and social performance in the apparel sector, and provides a framework for subsequent chapters and their case studies.

Methodology: a parsimonious approach to defining and measuring economic and social upgrading

Economic and social upgrading are multi-faceted and, thus, highly complex processes. Broadly speaking, economic upgrading can be defined as 'the process by which economic actors – firms and workers – move from low-value to relatively high-value activities in global production networks' (Gereffi 2005: 171). Social upgrading, in turn, can be conceived as a process of improvement in the entitlements and rights of workers as social actors, which enhances the quality of their employment through the advancement of decent work and respect for labour standards (Barrientos et al. 2011). These are very broad definitions that pay tribute to the complexity of the underlying processes.

However, in order to be able to measure these phenomena using available data, it is necessary to come up with narrower definitions that are easier to operationalize (see Milberg and Winkler 2011). The present chapter, therefore, applies a *parsimonious approach* to the study of economic and social upgrading in the apparel sector. It uses quantitative secondary data to trace upgrading (or downgrading) trajectories of a number of developing countries integrated in the global apparel value chain. Instead of focusing on detailed portrayals of individual supplier countries this approach allows us to sketch overall trends in the global apparel sector and to compare up/downgrading dynamics across countries and regions.

Economic upgrading

Following Kaplinsky and Readman (2005) and Amighini (2006), a country's apparel sector is said to experience *economic upgrading* when the following two conditions are fulfilled:

1) there is an increase in its world export market share, reflecting international competitiveness of its exports; and
2) there is an increase in the export unit value, implying the production of higher-value products.[3]

It is essential to look at *changes* in these indicators *over time* to capture the dynamic nature of *upgrading (or downgrading) as a process*. Moreover, it is important to include *both* dimensions in our analysis as they convey complementary information that allows us to more adequately capture economic upgrading.

Export unit values[4] are commonly used as surrogates for price indices and, consequently, as proxies for product quality (e.g. IMF et al. 2009). Aiginger (1997: 574) points out that '[a] country with a higher [export] unit value will in some sense supply more quality'. He suggests that export unit values can be used 'as a complementary indicator for assessing the qualitative side of competitiveness' (Aiginger 1998: 60). Various recent studies of GVCs suggested the use of export unit value data to capture economic upgrading (Evgeniev and Gereffi 2008; Frederick and Staritz 2012; and Staritz 2011).

However, it is important to note that increasing export unit values could, in principle, also reflect rising production costs, which over time would lead, in a competitive market, to a loss of international competitiveness. For this reason, it is crucial to use the change in export market shares as a second, complementary indicator in our analysis to identify instances of economic upgrading more accurately and to avoid misleading interpretations.

Social upgrading

Social upgrading is defined to occur in a given country's apparel sector when the following two conditions are fulfilled:

(1) there is an increase in sectoral employment; and
(2) there is an increase in sectoral real wages.

The rationale for choosing these two indicators of social upgrading is straightforward: through the creation of jobs, labour is given the possibility to earn income – which potentially helps workers in developing countries to move out of poverty and thereby contributes to social wellbeing.[5] Real wages, on the other hand, are a measure of how much workers benefit or appropriate from the value created by economic activity in their country's apparel sector. These quantitative indicators do not fully capture the qualitative features of social upgrading, but real wages are taken here as a proxy for the *quality* of employment as an indicator of social upgrading.

Country selection

This parsimonious approach was applied to identify economic and social upgraders and downgraders among a sample of developing countries (Table 2.1). The selection of countries was determined by the list of countries participating in the Better Work programme and country cases in the Capturing the Gains research network. These were supplemented by major economies for each continent; China, India, Mexico and South Africa. The remaining countries were included to reflect a balanced regional distribution and on the basis of their role as important apparel suppliers.

Table 2.1 Country sample

Region	Country
Africa	Kenya, Lesotho, Mauritius, South Africa
Asia and Middle East	Bangladesh, Cambodia, China, India, Indonesia, Jordan, Sri Lanka, Viet Nam
Central America and the Caribbean	Dominican Republic, El Salvador, Guatemala, Haiti, Mexico, Nicaragua

Economic upgrading and downgrading in the global apparel sector in the 2000s

With the increasing international fragmentation of production and the rise of GVCs developing countries' stakes in global apparel production have increased. Today, apparel production is organized in classic 'buyer-led' chains (Gereffi 1994), in which lead firms source globally. As a result, a number of developing countries have become major producers of apparel products, and several of them are now among the top apparel exporters.[6] Four of the top five (China, Bangladesh, Turkey and India) and ten of the top 15 apparel exporting countries in terms of world export market share are developing countries (including Viet Nam, Indonesia, Cambodia, Thailand, Mexico and Morocco (Table 2.2)). By comparison, the other (mainly African, Central American and Caribbean) countries in this sample are small apparel exporters, commanding market shares (sometimes well) below 1 per cent (see Table 2.3).

In the following, first data on the two economic upgrading indicators are presented and discussed before interesting patterns and trends are highlighted. The definition specified already is then applied to explore the extent to which economic upgrading has taken place in the

Table 2.2 Top 15 apparel exporters in 2010

Rank	Country	World export market share (%)	Export value (in million US$)
1	China	42.58	140,693
2	Bangladesh	5.03	16,633
3	Turkey	4.48	14,788
4	Italy	4.42	14,599
5	India	3.90	12,897
6	Viet Nam	3.36	11,088
7	Germany	2.81	9,287
8	Indonesia	2.40	7,916
9	France	1.98	6,530
10	Spain	1.41	4,650
11	Mexico	1.27	4,202
12	Cambodia	1.27	4,196
13	Netherlands	1.21	4,011
14	Thailand	1.15	3,797
15	Morocco	1.14	3,770
	World	**100.00%**	**330,398**

Source: Author's own calculation based on UN Comtrade data.

selected developing countries. In doing so, particular attention is paid to the structural break that occurred in the apparel sector at the end of 2004 with the elimination of the MFA quota system, which considerably re-shaped the international division of labour, reallocating apparel production and employment between countries and increasing the importance of other factors such as labour costs, lead times, flexibility in production, trade preferences and the availability of nonmanufacturing capabilities in the supplier base (with global buyers increasingly preferring full-package suppliers over simple assemblers) (Frederick and Staritz 2012).

The left-hand panel of Table 2.3 shows for each country in the sample the value of apparel exports in million US$ as well as its world export market share for the years 2000, 2004, 2005, 2009 and 2010; the right-hand panel gives the growth rates of these two variables pre-MFA phase-out (2000–2004), post-MFA phase-out (2004–2010) and over the entire decade of the 2000s. Broadly speaking, throughout the first decade of the 2000s, the Asian countries increased export value and market share, whereas most African and Latin American and Caribbean countries have seen decreases in market share and often even in export value.

The situation in African countries is complex. Over the entire first decade of the 2000s, Kenya and Lesotho increased their market shares. However, this increase in market share occurred before the MFA phase-out. Since the end of the MFA, both countries have lost about 50 per cent of the market shares they commanded in 2004. The African Growth and Opportunity Act (AGOA) was important in that after its inception in 2000 it granted African countries preferential access to the US market, which attracted quota-seeking foreign investors and initially spurred exports in the early 2000s (see also Pike and Godfrey, Chapter 8 in this volume). However, with the expiration of the MFA, apparel production in Kenya and Lesotho lost its quota advantage vis-à-vis exporters from Asia. Without the advantages of unused MFA quotas many 'MFA quota-hopping firms' withdrew – despite AGOA (Staritz 2011: 55). Meanwhile, the other two African countries in our sample (Mauritius and South Africa) lost market share throughout the 2000s, with the downward trend accelerating after the MFA phase-out.

Interestingly, in most countries the trend prior to the MFA phase-out also continued afterwards. The exceptions (i.e. where the expiration of the MFA caused a reversal in the previous trend) are Guatemala, Jordan, Kenya and Lesotho. Each was negatively affected by the MFA phase-out, as their market share gains in the early 2000s turned into losses post-2004. Indonesia also saw a reversal, but in this case declines in market

Table 2.3 Apparel exports (in million US$) and world export market shares (%), 2000–2010

Country	2000	2004	2005	2009	2010	%-change 2000–2004	%-change 2004–2010	%-change 2000–2010
Bangladesh	4,862.64	7,956.99	8,041.71	14,252.64	16,632.82	63.64	109.03	242.05
Market share	*2.50*	*3.15*	*2.98*	*4.70*	*5.04*	*26.23*	*59.79*	*101.70*
Cambodia	1,215.12	2,435.78	2,699.84	3,489.22	4,195.39	100.46	72.24	245.27
Market share	*0.62*	*0.97*	*1.00*	*1.15*	*1.27*	*54.63*	*31.66*	*103.60*
China	48,065.05	71,240.18	89,933.26	124,555.57	140,539.12	48.22	97.28	192.39
Market share	*24.70*	*28.24*	*33.38*	*41.05*	*42.58*	*14.34*	*50.80*	*72.42*
Dominican Republic	2,480.84	2,113.70	1,908.20	670.55	675.86	–14.80	–68.02	–72.76
Market share	*1.27*	*0.84*	*0.71*	*0.22*	*0.20*	*–34.28*	*–75.56*	*–83.94*
El Salvador	1,689.00	1,818.48	1,718.49	1,422.13	1,770.90	7.67	–2.62	4.85
Market share	*0.87*	*0.72*	*0.64*	*0.47*	*0.54*	*–16.95*	*–25.56*	*–38.17*
Guatemala	1,561.39	2,068.74	1,939.22	1,231.36	1,293.88	32.49	–37.46	–17.13
Market share	*0.80*	*0.82*	*0.72*	*0.41*	*0.39*	*2.21*	*–52.19*	*–51.13*
Haiti	267.68	358.08	459.88	572.37	578.24	33.77	61.48	116.02
Market share	*0.14*	*0.14*	*0.17*	*0.19*	*0.18*	*3.19*	*23.44*	*27.38*
India	5,133.91	7,308.36	9,482.59	11,945.00	12,895.70	42.35	76.45	151.19
Market share	*2.64*	*2.90*	*3.52*	*3.94*	*3.91*	*9.81*	*34.88*	*48.12*
Indonesia	4,689.85	5,298.07	5,690.19	7,182.95	7,913.87	12.97	49.37	68.74
Market share	*2.41*	*2.10*	*2.11*	*2.37*	*2.40*	*–12.85*	*14.18*	*–0.49*
Jordan	68.23	1,030.52	1,157.48	826.38	873.41	1,410.31	–15.25	1,180.05
Market share	*0.04*	*0.41*	*0.43*	*0.27*	*0.26*	*1,065.07*	*–35.21*	*654.82*

(*continued*)

Table 2.3 Continued

Country	2000	2004	2005	2009	2010	%-change	%-change	%-change
						2000–2004	2004–2010	2000–2010
Kenya	50.34	306.8	297.16	212.98	222.5	509.44	–27.48	341.97
Market share	*0.03*	*0.12*	*0.11*	*0.07*	*0.07*	*370.13*	*–44.56*	*160.62*
Lesotho	152.51	494.27	421.7	303.47	317.74	224.08	–35.71	108.34
Market share	*0.08*	*0.20*	*0.16*	*0.10*	*0.10*	*150.00*	*–50.86*	*22.85*
Mauritius	961.11	958.52	807.37	817.51	769.25	–0.27	–19.75	–19.96
Market share	*0.49*	*0.38*	*0.30*	*0.27*	*0.23*	*–23.07*	*–38.65*	*–52.80*
Mexico	8,925.97	7,286.73	6,684.33	3,928.56	4,201.98	–18.36	–42.33	–52.92
Market share	*4.59*	*2.89*	*2.48*	*1.29*	*1.27*	*–37.03*	*–55.92*	*–72.24*
Nicaragua	351.6	624.16	753.55	960.33	1,096.20	77.52	75.63	211.78
Market share	*0.18*	*0.25*	*0.28*	*0.32*	*0.33*	*36.94*	*34.25*	*83.85*
South Africa	396	476.66	335.12	166.32	177.03	20.37	–62.86	–55.29
Market share	*0.20*	*0.19*	*0.12*	*0.05*	*0.05*	*–7.15*	*–71.61*	*–73.64*
Sri Lanka	2,519.67	2,975.61	3,084.93	3,541.67	3,732.82	18.10	25.45	48.15
Market share	*1.29*	*1.18*	*1.14*	*1.17*	*1.13*	*–8.90*	*–4.11*	*–12.64*
Viet Nam	1,684.50	4,493.65	4,831.45	9,523.39	11,088.37	166.76	146.76	558.26
Market share	*0.87*	*1.78*	*1.79*	*3.14*	*3.36*	*105.79*	*88.63*	*288.16*

Source: Author's own calculation based on UN Comtrade data.

shares in the first years of the 2000s were re-gained afterwards. In the case of Jordan, the success of the early 2000s was driven by preferential access to the US market granted in the framework of Qualifying Industrial Zones established in 1996 (see Pickles 2012b). However, part of this competitive advantage eroded with the demise of the MFA.

Between 2000 and 2010 the Dominican Republic, Guatemala, Mexico, Mauritius and South Africa not only lost market shares but even saw their export values decline, whereas all the other countries in the sample managed to increase the value of their exports (albeit not all of them at the same pace of growth as the world apparel market so that their respective market share declined, e.g. El Salvador or Sri Lanka).

Evidence for the second economic upgrading indicator, the growth of export unit values,[7] suggests that progress has been more widespread. Except for some Central American and Caribbean economies (the Dominican Republic, El Salvador, Haiti and Nicaragua), all countries in the sample saw the unit value of their exports (measured as price per kilogram) increase between 2000 and 2010 (Table 2.4).

However, this overall finding masks some interesting sub-trends. In fact, with the exception of China, the Dominican Republic and all the African countries in the sample, export unit values went down everywhere in the first part of the 2000s (i.e. pre-MFA expiration). Also, except for China and Mauritius (+29.2 and +16.1 per cent between 2000 and 2004, respectively), apparel export unit values did not keep pace with the growth rate of the world average (+15.7 per cent). By contrast, in the second half of the first decade of the 2000s all countries in the sample except the Dominican Republic, Kenya and Nicaragua increased their export unit values, a surprising finding given that it was widely expected that the MFA phase-out would lead to an intensification of international competition and, consequently, to a downward pressure on prices in the global apparel sector. However, looking at the entire decade, apparel export unit values grew faster than the world average only in China, Mauritius and South Africa (see the last column in Table 2.4).

For the post-MFA period, Figure 2.1 combines the two economic upgrading indicators in a 2 × 2 matrix. The countries in the upper right-hand quadrant are considered clear-cut economic upgraders (according to the definition presented already); they have managed to couple export market share gains (plotted on the horizontal axis) with increases in their export unit values (plotted on the vertical axis). By contrast, the countries that appear in the lower left-hand quadrant are considered clear-cut economic downgraders, with declines in both their export market shares and their export unit values between 2004 and 2009.

Table 2.4 Aggregated apparel export unit values (in US$/kilogram)

Country	**2000**	**2004**	**2005**	**2009**	**2010**	**%-change**	**%-change**	**%-change**
						2000–2004	**2004–2010**	**2000–2010**
Bangladesh	12.52	12.09	11.64	15.26	15.42	–3.43	27.54	23.16
Cambodia	16.91	16.63	16.36	17.86	18.18	–1.66	9.32	7.51
China	12.35	15.95	14.53	18.54	18.54	29.15	16.24	50.12
Dominican Republic	16.52	16.68	16.05	16.87	14.96	0.97	–10.31	–9.44
El Salvador	12.33	10.87	10.37	11.84	11.46	–11.84	5.43	–7.06
Guatemala	16.84	14.67	14.78	16.77	17.26	–12.89	17.66	2.49
Haiti	10.79	7.58	6.9	7.64	8.44	–29.75	11.35	–21.78
India	19.25	17	18.3	21.3	21.7	–11.69	27.65	12.73
Indonesia	18.57	17.8	17.02	21.34	20.43	–4.15	14.78	10.02
Jordan	18.36	16	15.95	17.68	18.94	–12.85	18.38	3.16
Kenya	11.89	12.97	12.34	12.47	12.2	9.08	–5.94	2.61
Lesotho	12.29	12.55	13.2	12.63	13.35	2.12	6.37	8.62
Mauritius	19.57	22.72	23.15	29.95	30.65	16.10	34.90	56.62
Mexico	12.34	12.29	12.59	13.52	13.72	–0.41	11.64	11.18
Nicaragua	13.28	13.2	11.51	11.14	11.26	–0.60	–14.70	–15.21
South Africa	11.7	13.25	15.32	17.68	n.a.	13.25	33.43	51.11
Sri Lanka	21.03	19.41	20.72	25.12	24.91	–7.70	28.34	18.45
Viet Nam	16.87	16.38	16.85	20.06	20.45	–2.90	21.36	21.22
World average	**15.76**	**18.23**	**16.97**	**19.81**	**20.39**	**15.67**	**11.85**	**29.38**

Source: Author's calculation based on UN Comtrade data.

Figure 2.1 indicates that economic upgrading after the MFA phase-out has been concentrated in Asia; six of the seven countries that qualify as clear economic upgraders are Asian, the only exception being Haiti whose success is directly related to its duty-free access to the US market under the HELP and HOPE I and II programmes (Frederick and Gereffi 2011; Pickles 2012b). For some Asian exporters (Bangladesh, Cambodia and Viet Nam), this positive development after the MFA phase-out has been quite surprising given that quota-hopping had contributed considerably to the development of their apparel industries so that they were expected to lose out significantly with the dismantling of the MFA (Gouvea Abras 2012; Staritz 2011). However, all three countries managed effectively to turn their low labour costs into a genuine competitive advantage and to utilize the preferential access they have been enjoying, particularly to the EU market. Moreover, in Bangladesh and Viet Nam, proactive government policies to prepare for the MFA phase-out and to encourage industry upgrading have played a crucial role for the strong performance of their apparel firms. Both countries also saw considerable investments in backward linkages into the textile industry, which helped

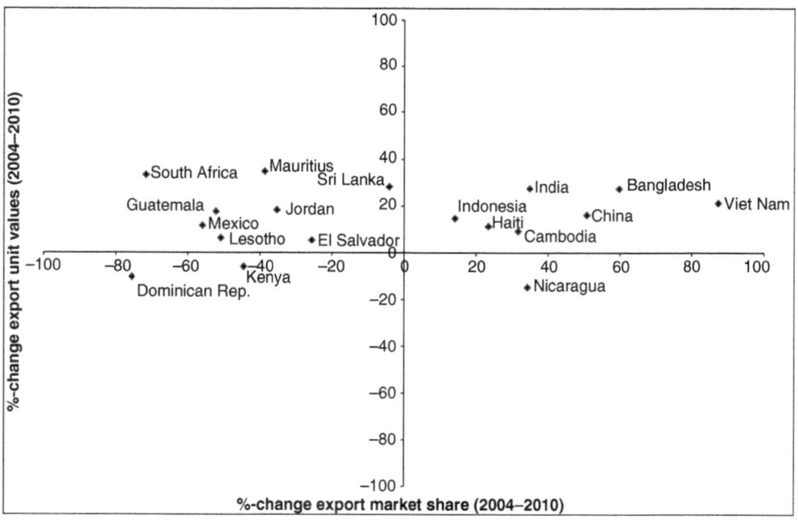

Figure 2.1 Economic upgrading and downgrading in the apparel sector, 2004–2009

Note: In order to even out year-on-year fluctuations and to minimize the impact of data outliers, we used three-year averages, that is we calculated the percentage change from 2003–2004 to 2008–2010.

Source: Author's illustration based on data from UN Comtrade.

to improve supply chains, delivery time and competitiveness. In the case of Bangladesh, in particular, for a part of the sector this has facilitated functional upgrading from assembly to full-package apparel production, including the provision of sourcing and logistics services (see Staritz and Frederick 2012). In Cambodia, while government support was limited and product and functional upgrading rather scarce, trade preferences, the ILO's Better Factories Cambodia programme and the reputation for ethical sourcing played an important role in the survival and even success of the apparel industry post-MFA (Brown et al. 2011; Staritz and Frederick 2012).

For other countries, particularly China and India, the positive post-MFA performance was less surprising given that they had faced quota restrictions during the MFA (Thoburn 2010). Frederick and Gereffi (2011) describe China as the big winner in the global apparel sector since 2000 and list a number of explanatory factors for its success: government support, strong backward linkages, heavy investment in machinery leading to process upgrading and productivity growth, the availability of skilled and cheap labour and management capabilities, a diversified export product basket and the diversification of end markets into the dynamic domestic market and emerging regional markets. The positive performance of India's apparel sector can also largely be explained by its backward linkages to a well-established fibre-textile supply chain and its increasingly dynamic domestic market, both of which have fostered the development of a broad range of functional capabilities. This functional upgrading has been important in bolstering the competitiveness of India's apparel sector after the MFA, especially given its comparatively more sophisticated – and higher-priced – export basket (as reflected in its export unit values, see Table 2.4). Targeted domestic policies and local entrepreneurship also contributed decisively to the prosperity of India's apparel sector post-MFA (Frederick and Staritz 2012; Tewari 2008).

By contrast, the Dominican Republic and Kenya appear to be the only outright economic downgraders in the sample. As mentioned already, Kenya's apparel sector suffered from a reversal of 'quota-hopping' after the MFA phase-out. The Dominican Republic, on the other hand, as one of the regional suppliers to the United States whose key competitive advantage is underwritten by the Dominican Republic–Central America Free Trade Agreement (CAFTA-DR) preferential market access and proximity to the end market, has paid for its dependence on demand from the United States. Moreover, it has suffered from some of the drawbacks of CAFTA-DR, which offers preferential access to the US market, but only for apparel products assembled under complex rules of origin that favour US-made fabric (Pickles 2012). This reliance on more expensive US inputs has harmed the international competitiveness while US trade preferences have augmented competition among regional suppliers (Bair and Gereffi; Frederick and Gereffi 2011, Chapter 11 in this volume).

The remaining two quadrants in Figure 2.1 reflect the 'intermediate cases' where progress on one indicator was accompanied by regression on the other. Half of the countries in the sample rank as 'intermediate cases'. Within this group, the bulk of countries registered declines in their market shares but increases in their export unit values (upper left-hand quadrant). One could conjecture that in these cases price increases went at the expense of international competitiveness, as reflected in declining market shares. However, the group of countries in this quadrant is quite heterogeneous and their underlying stories differ. For example, in South Africa apparel production is increasingly geared towards the domestic market (Morris and Barnes 2009: 42), which is reflected in a reduction in both export values and export market share. However, those articles that are still produced for export are in a higher-value segment, which, together with rising wages (see Table 2.6), explains the rise in export unit values. Sri Lanka is a prominent case of both functional and product upgrading where production has shifted from lower-value to more complex higher-value products, especially lingerie articles (Staritz and Frederick 2012), which is mirrored in the increase in export unit values. At the same time, this moving out of lower-value segments resulted in a loss of world market shares. The story of Mauritius is quite similar. Many foreign-owned firms that had been serving the US market left Mauritius when the MFA expired while locally owned and European firms, which offered more capabilities and produced higher-value products, expanded their exports to the EU market (Staritz 2011). Overall, this resulted in a decline of world market shares and an increase in export unit values.

In the case of Mexico, the increase in export unit values does not so much reflect product upgrading but primarily rising production costs. Labour costs are relatively high (see Table 2.6) and the costs of inputs – with yarns and fabrics being largely imported from the United States within production-sharing arrangements – have increased. Limited government support, the lack of broad upgrading among suppliers (who mostly operate under the traditional *maquiladora* model) and the failure to diversify export markets beyond the United States have further contributed to Mexico's slipping market position in a context of increased competition from low-cost Asian suppliers (Staritz and Frederick 2012).

Lesotho's apparel manufacturing has also not upgraded much post-MFA so that, despite continued concentration on low-value assembly activities, slow productivity growth combined with a lack of backward linkages and infrastructural capacities (which contributed to rising costs) have resulted in rising export unit values. However, at the same time, more suppliers are integrated in newly emerging South African-owned supply chains that produce comparatively more complex and higher-value articles for export to South Africa, which has also contributed to higher aggregate unit values of Lesotho's exports (Morris et al. 2011; Pike and Godfrey; Staritz 2011, Chapter 8 in this volume).

The only intermediate case in the sample that appears in the lower right-hand quadrant of Figure 2.1 is Nicaragua, which combined decreasing export unit values with market share gains. Nicaraguan garment producers have had limited success in upgrading and in moving up the value chain. In fact, they are still mainly engaged in assembly services and the expansion of exports was mainly in low-value articles. The drop in Nicaragua's unit values, thus, has to do with a shift in the composition of its export basket in favour of cheaper apparel items. Yet, in these low-value segments, Nicaragua has managed to stay competitive and even increase its market share post-MFA largely thanks to preferential access to the US market within the CAFTA-DR (Bair and Gereffi, Chapter 11 in this volume; Fernandez-Stark et al. 2011).

To sum up, there has been some important post-MFA economic upgrading in the sample countries. Seven of the 18 countries studied managed to gain export market shares and at the same time also increase the unit values of their apparel exports. These countries appear to have enhanced their competitiveness and, thus, qualify as unequivocal economic upgraders according to our definition. With the exception of Haiti, all of these post-MFA winners are Asian countries. Interestingly, only two apparel exporters in the sample, the Dominican Republic and Kenya, experienced clear-cut economic downgrading after the MFA phase-out, reflecting the fact that not even a decline in their export unit values could prevent their losses of market shares. Meanwhile, half of the developing countries studied experienced ambiguous progress, with improvements on one front but deterioration on the other. Most of these 'intermediate cases' lost export market shares while increasing their export unit values – the latter reflecting product and/or functional upgrading in some cases, and rising production costs and loss of competitiveness in other cases. Among the African, Central American and Caribbean countries in the sample, almost all are classified either as intermediate cases or as economic downgraders. Overall, the findings thus confirm the pattern also identified in the case study literature that, in general, regional suppliers have lost out to the low-cost and often more capable apparel producers in Asia.

Social upgrading and downgrading experiences

In a second step, we now look at how the countries in the sample fared in terms of social upgrading. In this context, data gaps are much more of an issue than for the economic upgrading indicators. The statistical recording of information on employment and wages, particularly at a sufficiently detailed sectoral level, is not as comprehensive as for trade flows so that for some countries and years data are scarce or not available at all. The categorization of economic activities can differ somewhat between data sources so that the definition of what constitutes the 'apparel sector' also varies across countries and/or indicators.[8] The time spans of data coverage are not always identical across the countries in the sample. Such data unevenness limits the comparability of the social upgrading analysis, first, with the findings from the economic analysis and, second, across countries. As a result, inferences made here should be interpreted with some caution. The same caveat applies to the analysis on the relationship between economic and social upgrading undertaken in the 'overall upgrading' section.

The first indicator of social upgrading (change in apparel employment) suggests a clear regional pattern (Table 2.5). In all but one of the Asian countries in our sample, the number of jobs in the apparel sector increased throughout the first decade of the 2000s (both before and after the MFA phase-out). The exception was Sri Lanka where employment increased in the early 2000s but fell after 2004. In the cases of Bangladesh and, in particular, Viet Nam, these rapid rates of employment expansion reflect the burgeoning demand for labour that accompanied the increasing integration of these two countries into international apparel value chains (see export growth rates in Table 2.3).

In China and India, rising incomes and the emergence of a middle class have contributed to an increasing importance of the domestic market, which has helped to maintain or even further spur employment generation (Frederick and Gereffi 2011).[9] This re-orientation towards domestic consumers notwithstanding, the recent global economic crisis did not leave the apparel sector in China and India unscathed and employment has declined slightly since 2008 (Forstater 2010). In India the informal sector is large (Gouvea Abras 2012: 88) and has absorbed some of the labour from lost formal sector jobs during the

Table 2.5 Employment in the apparel sector, 2000–2009

Country	2000	2004	2005	2008	2009	%-change		
						2000–2004	2004–2009	2000–2009
Bangladesh	1,600,000	2,000,000	2,000,000	2,800,000	3,100,000	25.00	55.00	93.75
Cambodia	168,824	269,846	283,906	324,871	281,855	59.84	4.45	66.95
China	2,156,300	3,202,600	3,460,600	4,587,000	4,493,100	48,52	40,30	108,37
Dominican Republic	141,945	131,978	91,491	49,735	41,285	-7.02	-68.72	-70.91
El Salvador	131,300	123,300	105,400	107,100	n.a.	-6.09	-13.14	-18.43
Guatemala	88,255	113,272	87,682	n.a.	56,702	28.35	-49.94	-35.75
Haiti	n.a.	n.a.	n.a.	n.a.	n.a.	n.a.	n.a.	n.a.
India	329,401	447,466	538,615	622,913	n.a.	35.84	39.21	89.10
Indonesia	479,155	438,045	451,938	495,192	464,465	-8.58	6.03	-3.07
Jordan	14,216	18,002	18,427	22,410	29,460	26.63	63.65	107.23
Kenya	25,288	34,614	34,234	25,766	24,359	36.88	-29.63	-3.67
Lesotho	16,866	47,998	37,608	41,753	33,742	184.58	-29.70	100.06
Mauritius	72,810	59,691	52,659	50,924	42,355	-18.02	-29.04	-41.83
Mexico	640,000	482,396	409,910	314,343	289,351	-24.63	-40.02	-54.79
Nicaragua	32,220	40,940	56,335	50,712	51,850	27.06	26.65	60.92
South Africa	124,001	99,558	76,792	55,892	49,698	-19.71	-50.08	-59.92
Sri Lanka	280,000	306,984	273,600	270,000	280,000	9.64	-8.79	0.00
Viet Nam	231,948	498,226	511,278	758,274	n.a.	114.80	52.19	226.92

Note: For the Dominican Republic, Kenya and Nicaragua, data refer to employment in export processing zones (EPZs). For Lesotho, figures are for employment in the textiles and clothing sector. For El Salvador, the figure for 2008 refers to 2007; for Guatemala, the figure for 2000 refers to 2001; for Kenya and Nicaragua, the figure for 2000 refers to 2002; and for Sri Lanka, the figure for 2000 refers to 1999. The growth rates reported in the last three columns cover different time periods accordingly.

Sources: Author's illustration based on data from UNIDO INDSTAT4 and ILO LABORSTA databases, National Bureau of Statistics of China, ASIES (2007, 2010), CNZFE (2005, 2010), EPZA (2009), Lesotho Bureau of Statistics (2010), ProNicaragua (2011), Staritz and Frederick (2012).

crisis. Nicaragua is another success story as apparel producers in export processing zones (EPZs) generated almost 20,000 new jobs between 2002 and 2009.

Indonesian workers seem to have benefited from the MFA phase-out: apparel employment decreased between 2000 and 2004, but increased after 2004 and before the declines resulting from the economic crisis of 2008.

In the case of Jordan, it is important to qualify the impressive rise in employment as many of the jobs created have been filled by temporary migrant workers from South Asia who are estimated to account for 70–75 per cent of all employment in the Jordanian apparel sector, raising questions about the real extent of employment creation for the Jordanian economy (Brown and Deardoff 2011).

By contrast, among the poorest employment performers in the sample are the African and most Central American and Caribbean countries where apparel job loss in the first decade of the 2000s was significant. Labour was hit hardest in the Dominican Republic, Mexico and South Africa where employment declined through the entire decade – with quite an acceleration after the MFA phase-out. By 2009 over half of all jobs had been cut back. In El Salvador and Mauritius, apparel employment also declined since the beginning of the new millennium, yet at a more moderate speed (–18 and –42 per cent, respectively). In Guatemala, Kenya and Lesotho the apparel sector added jobs in the early 2000s but when the MFA expired in 2004 this upward trend was reversed.

Table 2.6 presents data on labour earnings in the apparel sector (the second indicator for social upgrading). It is difficult to find consistent information across countries, which is why we have to settle for different indicators or different units across countries. The top panel of Table 2.6 shows the average nominal annual wages[10] in US$ for various countries for the years 2000, 2004, 2005, 2008 and 2009.[11] For other countries no wage data could be found. In these cases data on hourly labour costs (including social charges) are used instead, as gathered by Jassin-O'Rourke Group (2002, 2008), as a proxy for labour earnings (see bottom panel of Table 2.6). For Nicaragua, due to the lack of availability of data on actual wages, the table indicates the minimum annual wages in US$ (including social charges) that are set through tripartite agreements (among trade unions, employers and the government). The last three columns present the growth rates of nominal wages pre-MFA phase-out (2000–2004), post-MFA phase-out (2004–2009) and over the full first decade of the 2000s.

What is of relevance for our analysis is how wages in the apparel sector have developed over the last decade. The most remarkable cases in this regard are China (where nominal wages almost tripled between 2003 and 2009) and Nicaragua, India and Jordan (where wages more than doubled over the first decade of the 2000s). In China, this increase in wages in US$ terms was driven by a tightening of labour legislation, currency appreciation and rapid economic growth creating labour shortages (which strengthen workers' bargaining power). China was able to preserve its international competitiveness thanks to high labour productivity, supply chain development and backward linkages, and well-developed infrastructure systems (Frederick and Gereffi 2011). As for the positive development of wages in the Nicaraguan apparel sector, the strong industrial relations system and the fairly well-established social dialogue, as incarnated in tripartite negotiations (with a relatively active role of the Ministry of Labour), have played important roles. Yet, labour productivity remains relatively low in Nicaragua. However, comparatively good infrastructure, together with preferential market access to the US and favourable exchange rate developments,[12] has helped Nicaragua to stay competitive, particularly in low-value segments (MSN/EMIH/PASE 2012).

Other countries where apparel workers saw wage increases include Indonesia and South Africa (+77 and +61 per cent from 2000 to 2009, respectively). Labour income grew more slowly in Cambodia, Guatemala, Haiti, Lesotho and Mexico where wages in the later part of the first decade of the 2000s were less than 10 per cent higher than they were in the early 2000s. In Bangladesh, the Dominican Republic and Sri Lanka nominal wages or labour costs in US$ terms declined between 2000 and 2010. In Cambodia, apparel wages fell slightly in the early 2000s but increased after 2004. This upward trend in the second half of the decade can be explained, at least in part, by the increase in minimum wages that was agreed in tripartite negotiations in 2006 (see Arnold, Chapter 9 in this volume) and the positive effects of the Better Factories Cambodia programme (Robertson 2011).

Kenya and Lesotho experienced the opposite trend in wages after the 2004 MFA phase-out. In both countries apparel wages in US$ terms rose until 2004 but then started to fluctuate so that average earnings in 2009 were below the 2004 level. In Kenya this was primarily due to a dramatic crisis-related drop in wages in 2009, with higher wages in both 2008 and 2010. For Lesotho the picture is more complicated, with even more pronounced oscillations in average wages. Exchange rate fluctuations were important: in fact, in local currency terms average

Table 2.6 Nominal wages and labour costs in the apparel sector (in US$), 2000–2009

Country	Indicator / Unit	2000	2004	2005	2008	2009	%-change		
							2000–2004	2004–2009	2000–2009
Cambodia	Avge annual wage in US$	753	705	n.a.	888	834	–6.25	18.16	10.77
China	Avge annual wage in US$	1,256	1,402	1,578	3,094	3,661	n.a.	161.09	191.46
Dominican Republic	Avge annual wage in US$	3,718	2,182	3,380	3,466	3,537	–41.32	62.10	–4.87
El Salvador	Avge annual wage in US$	1,988	2,084	2,720	2,853	n.a.	4.81	36.88	43.46
India	Avge annual wage in US$	777	1,032	1,136	1,642	n.a.	32.80	59.16	111.36
Indonesia	Avge annual wage in US$	752	1,127	1,012	1,370	1,330	49.79	18.07	76.85
Jordan	Avge annual wage in US$	1,737	1,947	2,275	4,107	3,444	12.09	76.87	98.26
Kenya	Avge annual wage in US$	747	987	938	1,139	926	32.10	–6.15	23.97
Lesotho	Avge annual wage in US$	1,271	1,656	1,811	1,142	1,325	30.28	–19.98	4.25
Mauritius	Avge annual wage in US$	2,833	3,562	3,645	n.a.	3,577	25.73	0.43	26.26
Mexico	Avge annual wage in US$	4,829	4,950	5,467	6,312	5,328	2.51	7.64	10.34
South Africa	Avge annual wage in US$	3,758	5,161	6,350	5,718	6,066	37.35	17.53	61.43
Nicaragua	Min. annual wage in US$, incl. social charges	963	1,243	1,362	2,138	2,202	29.10	77.12	128.66
Bangladesh	Labour costs (US$ per hour, incl. social charges)	0.39	n.a.	0.23	0.22	n.a.	–41.03	–4.35	–43.59

(continued)

Table 2.6 Continued

Country	Indicator / Unit	2000	2004	2005	2008	2009	%-change		
							2000–2004	2004–2009	2000–2009
Guatemala	LC (US$/hour, incl. SC)	1.49	n.a.	n.a.	1.65	n.a.	n.a.	n.a.	10.74
Haiti	LC (US$/hour, incl. SC)	0.49	n.a.	n.a.	0.52	n.a.	n.a.	n.a.	6.12
Sri Lanka	LC (US$/hour, incl. SC)	0.48	0.46	n.a.	0.43	n.a.	–4.17	–6.52	–10.42
Viet Nam	LC (US$/hour, incl. SC)	n.a.	n.a.	0.28	0.38	n.a.	n.a.	35.71	n.a.

Notes: 'LC (US$/hour, incl. SC)' stands for 'Labour costs (US$ per hour, including social charges)'. For Cambodia and Lesotho, the figures refer to the textiles and clothing sector. For the Dominican Republic, Kenya and Nicaragua, the figures reported are wages in textiles and apparel in EPZs. For El Salvador, the figure for 2000 refers to 2001; for Bangladesh, Cambodia, Guatemala, Haiti, Kenya, and Sri Lanka, the figure for 2000 refers to 2002; for China and Mexico, the figure for 2000 refers to 2003; for Mauritius, the figure for 2009 refers to 2007. For El Salvador, India, Bangladesh, Guatemala, Haiti, Sri Lanka and Viet Nam, the final year underlying the calculation of growth rates is 2008. The growth rates reported in the last three columns cover different time periods accordingly.

Sources: Author's own illustration based on wage data from UNIDO INDSTAT4 database, Jassin-O'Rourke Group (2002, 2008), CNZFE (2005, 2010), EPZA (2009), Lesotho Bureau of Statistics (2010), Instituto Nacional de Estadística, Geografía e Informática (INEGI) de México, ProNicaragua (2011), and exchange rate data from IMF International Financial Statistics (IFS) database.

earnings in 2009 exceeded their 2004 level, reflecting a slight upward trend in nominal wages in the apparel sector over the ten years. Only when converted into US$ average earnings in 2009 are they below their 2004 level.

Combining the data on changes in employment and real wages to assess social upgrading or downgrading suggests that clear-cut social *upgrading* in the apparel sector was rare after the MFA phase-out. Only China, India, Jordan and Nicaragua were unambiguous social upgraders (upper right-hand quadrant of Figure 2.2) with both employment and real wages growth during 2004–2009.

Social *downgrading*, on the other hand, was widespread, with six sample countries registering a decline in both employment and real wages in apparel. All African countries in the sample (Kenya, Lesotho, Mauritius and South Africa) along with Guatemala and Sri Lanka were unambiguous downgraders (bottom left-hand quadrant of Figure 2.2). While employment in the Sri Lankan garment sector declined only slightly, real wages saw the most dramatic decline of all countries in the sample.

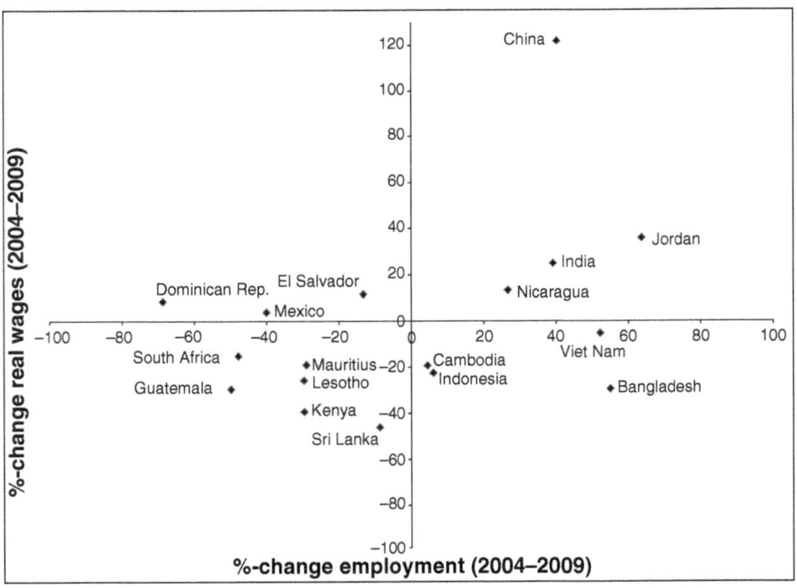

Figure 2.2 Social upgrading and downgrading in the apparel sector, 2004–2009
Note: Time spans covered are different for Bangladesh (2005–2008), El Salvador (2004–2008), Guatemala (2002–2008) and Viet Nam (2005–2008).
Source: Author's own illustration; data sources as indicated for Tables 2.5 and 2.6.

Seven countries in the sample figure as 'intermediate cases' (with opposite developments on the two indicators), reveal an interesting regional pattern. All the countries in the upper left-hand quadrant of Figure 2.2 (employment decline with real wage increase) are Central American and Caribbean, while all the countries in the lower right-hand quadrant are Asian (employment gain with real wage losses). The latter include many of the recent economic success stories, namely Bangladesh, Cambodia, Indonesia and Viet Nam. In these cases, growing demand for labour (triggered by increases in apparel production and exports) was accompanied (and perhaps driven by) declining real wages.

Of the countries located in the upper left-hand quadrant, El Salvador and Mexico also show up in the same quadrant in the economic upgrading matrix (see Figure 2.1); in these cases export market share losses (and indeed export *value* losses) were accompanied by job cuts, while real wages and export unit values both increased. The Dominican Republic is the country in the sample that experienced the biggest losses in both export market shares and apparel employment.

Overall, the findings from the country sample suggest that social upgrading has not been easy to achieve in the apparel sector after the MFA phase-out, with many cases of mounting pressure on labour (as a cost factor). In fact, most of the countries studied here experienced either outright social downgrading or mixed successes, with improvements on one front but deterioration on the other. Compared to the patterns for economic upgrading, it seems that social upgrading has been more tenuous and social downgrading has been far more common.

'Overall upgrading': where did economic and social upgrading concur?

In this section, the parsimonious approach is applied to assess the relationship between economic performance and social performance. The derived single (composite) index of economic upgrading and the single (composite) index of social upgrading have been plotted against each other in the 2 x 2 matrix in Figure 2.3. Of the four possible scenarios depicted in Figure 2.3, the upper right-hand and the lower left-hand quadrants represent the clear-cut cases. Apparel exporters that combine economic upgrading and social upgrading for 'overall upgrading' appear in the upper right-hand quadrant. The lower left-hand quadrant, on the other hand, represents 'overall downgraders': those countries that have registered both economic and social downgrading. Countries located in

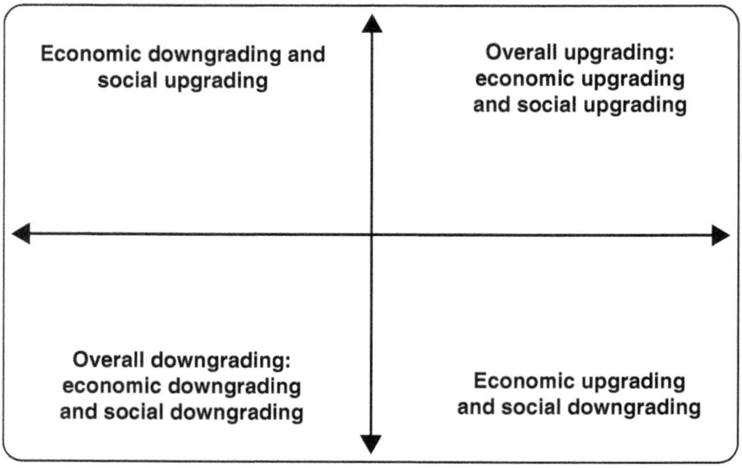

Figure 2.3 Prototype matrix of 'overall upgrading/downgrading'

the remaining two quadrants are intermediate cases, with upgrading on one front (either economic or social), but downgrading on the other.

To generate such matrices and, more generally, to relate changes in the economic and social realms to each other, it is necessary to create a single variable for each realm. This is indispensible as we have to reduce four dimensions (our two indicators for economic upgrading plus our two indicators for social upgrading) to just two dimensions (i.e. economic upgrading on the one hand and social upgrading on the other hand).

'Overall upgrading' calculations

To generate aggregate economic and social indices, all four underlying indicators are given equal weight to create the composite index symmetrically. To compute the composite indicator for 'economic upgrading', for example, a weight of 50 per cent each is assigned to both the percentage change in export market share and the percentage change in export unit value. The underlying formulas for the calculation of upgrading/downgrading are:

$$\textit{Economic up/downgrading} = 0.5 * (\textit{\%-change in market share}) + 0.5 * (\textit{\%-change in export unit value})$$

$$\textit{Social up/downgrading} = 0.5 * (\textit{\%-change in employment}) + 0.5 * (\textit{\%-change in real wages})$$

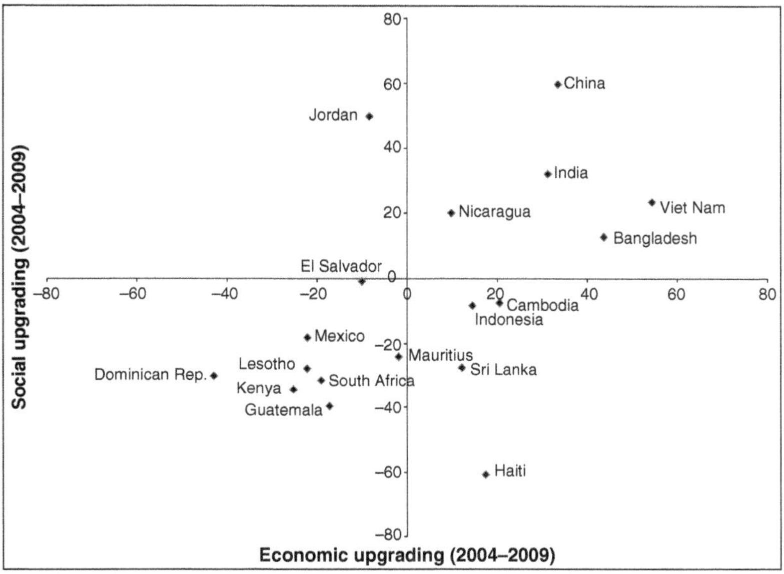

Figure 2.4 'Overall upgrading and downgrading' in the apparel sector, 2004–2009
Note: For Haiti, the measure of 'social upgrading' is based solely on the percentage change in real wages.
Source: Author's own illustration; data sources as indicated for Figures 2.1 and 2.2.

Based on this algorithm, Figure 2.4 shows the 2 × 2 matrix for 'overall up/downgrading'. It reveals that there have been quite some cases of overall upgrading after the MFA phase-out. Five of the 18 countries in the sample have experienced both economic and social upgrading (upper right-hand quadrant). Except for Nicaragua, this group of overall upgraders are all Asian apparel producers: Bangladesh, China, India and Viet Nam. Viet Nam and Bangladesh had the largest levels of economic upgrading, and China has been the prime performer in social upgrading. By contrast, overall downgrading has been a more widespread experience, affecting almost half of the apparel-producing countries in the sample. The Dominican Republic, Guatemala, Kenya, Lesotho and South Africa have been the worst performers, with substantial economic and social downgrading. In El Salvador, Mauritius and Mexico downgrading has been less pronounced. The remaining five countries in the sample are intermediate cases. Most have experienced economic upgrading and social downgrading (lower right-hand quadrant). This group includes the Asian exporters Cambodia, Indonesia and Sri Lanka, which have all performed quite well in economic terms but have not been able to match this with progress in social terms. Haiti has also experienced economic upgrading, but this was accompanied by a devastating social performance. Jordan is the only country located in the upper left-hand quadrant with social upgrading and economic downgrading. Overall, Figure 2.4 suggests that there is a positive correlation between economic upgrading and downgrading and social upgrading and downgrading in the apparel sector of these countries.

Connecting economic and social upgrading

Does economic upgrading translate into social upgrading? While our framework does not allow for a direct test of this relation, the findings generated by applying this parsimonious approach to the country sample provide some broad support for the view that there has been a positive *correlation* between economic upgrading and social upgrading in the apparel sector. A first indication is provided by visual inspection: although no clear pattern emerges, Figure 2.4 suggests a direct relationship between economic and social performance with countries clustered in the upper right-hand and lower left-hand quadrants. Second, and most importantly, with the exception of Jordan there is not a single case in the sample where social upgrading occurred without economic upgrading. In other words, all other countries that experienced social upgrading also experienced economic upgrading. However, the reverse is not the case; not every country that experienced economic upgrading also registered social upgrading.

However, it has to be emphasized that all that these exercises can at best indicate is a *correlation* between developments in the economic and social spheres. They tell us nothing about the *direction of causality* between the two. Causality may plausibly run in either direction, depending on how a number of different mediating factors (including the governance structure of value chains in which suppliers are integrated, the degree of unionization and labour organization, government policies and labour legislation, the extent of local ownership and the availability of trade preferences, among others) play out, and there is empirical evidence on both sides. Flanagan (2005) finds a tight correlation between productivity growth and wage growth in the apparel sector in a large sample of developing countries between 1995 and 2000. On the other side, Brown et al. (2011) find that improvements in working conditions in Cambodian apparel firms participating in the Better Work programme have been associated with a higher probability of plant survival and improved performance in terms of productivity and exports, citing efficiency wage theory (according to which higher wages may be associated with higher labour productivity) as one possible explanation. Nicaragua might be another example for the causality running from social upgrading to economic upgrading: its strong industrial relations system, as typified in tripartite agreements where, amongst other things, minimum wages are set for a period of three years, is a key competitive advantage and attractive to investors and global buyers as it offers stability and predictability on labour costs (Fernandez-Stark et al. 2011, MSN/EMIH/PASE 2012).

Conclusions

The global apparel sector has gone through considerable restructuring in recent years, particularly after the expiration of the MFA in 2004. This has impinged on the opportunities for upgrading of developing countries supplying to the global apparel value chain.

The analysis undertaken in this chapter has shown that there have been not only winners (upgraders) but also losers (downgraders). In general, while economic upgrading has been fairly widespread among the countries in the sample here, social upgrading has been more difficult to achieve. More precisely, this chapter has identified seven clear-cut economic upgraders but only four social upgraders. Conversely, while economic downgrading also occurred social downgrading has been far more common, mainly because of stagnant real wages. Our analysis found only two clear-cut economic downgraders but six social downgraders. Comparing across regions, economic and social upgrading have been concentrated in Asia, whereas downgrading mainly affected African countries and, to a lesser extent, the Central American and Caribbean countries in the sample.

Apart from tracing the up/downgrading trajectories of individual supplier countries, this chapter has also tried to address the question of whether improved international trade competitiveness translates into social gains. Simply improving export competitiveness and upgrading says little about how the gains are distributed and, thus, how social welfare is affected. While most research presumes that economic upgrading leads directly to social upgrading, this connection has not been analysed in a systematic fashion and there is ample evidence that transmission is not so straightforward. Here, composite indices were created to investigate the relationship between developments in the economic sphere and developments in the social sphere. While prudence is warranted when interpreting the findings from this parsimonious approach, they do seem to suggest that there has been a certain degree of positive correlation between economic and social performances (although this says nothing about *causality* between the two). Economic and social *upgrading* occurred together in five countries, while economic and social *downgrading* coincided in eight countries. Put differently, only in five out of 18 cases did the economic and social performance not correspond. With only one exception social upgrading has not occurred in the absence of economic upgrading. However, since not every country that recorded economic upgrading also experienced social upgrading, this analysis suggests that economic upgrading does not automatically translate into, but seems to be conducive to social upgrading. These results have to be taken as preliminary findings. Further analysis is necessary to learn more about the conditions under which economic and social upgrading are positively correlated or even mutually reinforcing, and to advance our understanding of the direction of causality between economic and social change.

Notes

1. This chapter draws on earlier work presented in Bernhardt (2013) and Bernhardt and Milberg (2013) which was undertaken within the Capturing the Gains research network. The author gratefully acknowledges valuable comments from Gary Gereffi, Shane Godfrey, Malte Luebker, Will Milberg, John Pickles, Arianna Rossi and Cornelia Staritz. The author also wants to thank the Kenyan Export Processing Zone Authority and ProNicaragua for kindly sharing their data.
2. These dual crises and their consequences on the global apparel sector are analysed in Gereffi and Frederick (2010) and in Lopez-Acevedo and Robertson (2012).
3. Progress on these two indicators might be interpreted to reflect what in the recent GVC literature, and the typology developed therein, is known as 'product upgrading', 'functional upgrading' and maybe also 'process upgrading' (e.g. Humphrey 2004; Gereffi et al. 2005; UNIDO 2011).
4. Export unit values are calculated by dividing the total value of a country's exports (of a certain commodity or product group) in a given period by the quantity or volume of these exports and, therefore, measure the average monetary value of one unit of export.
5. However, informal and casual forms of work are widely used in the apparel GVC and particularly important in several countries. The problem is that official data typically do not cover the informal sector and do not sufficiently account for irregular employment like temporary or contractual work (where working conditions and pay are usually worse than in regular jobs). Given this lack of reliable data on irregular employment, we do not know whether social upgrading/downgrading is accompanied by a rise or fall of precarious jobs.

6. For the purpose of measuring economic upgrading through the two indicators mentioned already, the apparel sector is defined as being comprised of chapters 61 ('Articles of apparel and clothing accessories, knitted or crocheted') and 62 ('Articles of apparel and clothing accessories, not knitted or crocheted') of the Harmonized Commodity Description and Coding System (HS) of tariff nomenclature. Please note that the definition of the apparel sector differs somewhat for the social upgrading indicators. All trade data are taken from the UN Comtrade database.
7. As noted, export unit values are calculated by dividing the total value of a country's annual exports by the quantity or volume of these exports. A complication is that quantities are usually recorded at higher levels of product disaggregation. For this reason, the UN Comtrade database offers information on export quantities only at the 4-digit level of product disaggregation but not at the 2-digit level (on which our sector definition is based). The sector-wide (or 'aggregated') export unit values that are presented here, therefore, have to be computed through the aggregation of export unit values at the 4-digit level, applying a weighted average method. However, given certain consistency problems and data gaps, the data reported here, especially the absolute figures, should be interpreted with care.
8. In some cases, data are only available for the 'textiles and clothing sector' and we have to make do with these data and use them as proxy for the apparel sector more narrowly. In other cases, wage data could be found only for certain occupational groups within the apparel sector (like garment cutters or sewing-machine operators) so that the evolution of their wage rates were taken as a proxy for the sector as a whole.
9. With a few exceptions where they relate to EPZs, the employment figures reported here cover the entire apparel sector, not just employment for apparel production geared towards export markets. That is, official data make no distinction between employment related to apparel production for export demand (stemming from the country's insertion into GVCs) and production for domestic demand. We, therefore, have to take the developments in the apparel sector as a whole as a proxy for developments in export-related apparel production related to participation in global value chains. This is an issue given that the indicators for economic upgrading are more narrowly based on export data. The same caveat also applies to the wage data.
10. *Nominal* wages are presented and discussed because they allow cross-country comparisons, especially when looking at wages as a cost factor. In Figure 2.2, however, (changes in) *real* wages (i.e. nominal wages deflated by the Consumer Price Index, CPI) are used as they give an indication of the development of workers' purchasing power. Real wages are, thus, more revealing about the extent to which labour has been able to appropriate the value created in apparel production, i.e. about the distribution of the value generated – which is a better indicator of social upgrading.
11. In some cases, these annual figures had to be calculated from data reported in different time units, e.g. average weekly or monthly wages. All wages reported in local currency were converted into US$ using exchange rate data from the International Financial Statistics database of the IMF.
12. According to IMF data, the local currency depreciated by about a third vis-à-vis the US$ between 2004 and 2010.

3

How 'Fair' Are Wage Practices along the Supply Chain? A Global Assessment[1]

Daniel Vaughan-Whitehead

Introduction

A number of recent reports and press media have highlighted the wage problems found among suppliers or subcontractors of a number of global apparel brands.[2] This reality discovered at the end of the supply chain echoes the worrying wage trends also identified over the last decade at a global level – on poor real wage progression, the increase in low paid workers, the decreasing wage share and the rising of wage inequalities – and that may have been even further exacerbated by the current financial and economic crisis (ILO 2008a, 2010; Bernhardt, Chapter 2 in this volume). All this evidence is leading to increasing concerns about wages,[3] and to a willingness of a series of actors – governments, NGOs, suppliers and brands themselves – to seriously address the crisis of wages.

For this to happen, however, there is a need for developing proper tools that would help to identify accurately wage problems, and also provide possible remedies to solve them. The debate on wages in the supply chain both at theoretical and practical levels has focused so far either on legal compliance (with the minimum wage, working hours, non-payment, wage left and delayed payments) or on the payment of a living wage. By contrast, the present chapter is aimed at providing a new theoretical framework and also fresh empirical evidence on wage practices along the supply chain. 'Fair wages' are defined through a new approach, the 'fair wage approach', aimed at providing CSR actors with a coherent set of fair wage dimensions and indicators (Vaughan-Whitehead 2010). This approach is applied in a large-scale exercise carried out on wages in more than 100 suppliers in Asia that is then complemented by three case studies in China.

The conclusion outlines the policy implications and the potential role of the different actors in a combined effort to improve wage practices along the supply chain.

Fair wages: a multi-dimensional approach

To date the debate about wages along the supply chain has been dominated by purely compliance issues – to legal regulations such as the minimum wage and working time regulations – and then extended to living wage issues – notably through the pressure and campaigns by a number of non-governmental organizations (NGOs). While these different approaches take into account important wage facets, they cannot reflect the whole wage story within an enterprise and, thus, not fully capture its 'fair wage' performance.

The methodology proposed here adopts a comprehensive and multi-dimensional approach to wage issues: it would be difficult for a company to make progress in one dimension, while not taking care of the others. This led the authors to identify the various types of wage developments ('fair wage dimensions') that help define what a fair wage should be. On the basis of field work results on wages, a general definition, complemented by an extended definition of fair wages (Box 3.1), is proposed.

Box 3.1 Definition of fair wages

General definition

Fair wages refer to company practices that lead to sustainable wage developments.

Extended definition

Fair wages refer to wage levels, wage progression and wage-fixing mechanisms that provide a living wage floor for workers, while complying with national wage regulations (such as the minimum wage, payment of wages, overtime payments, provision of paid holidays and social insurance payments) and lead to balanced wage developments within the company (with regard to wage disparity, skills, individual and collective performance and adequate internal communication and collective bargaining on wage issues).

The fair wage dimensions

The concept of fair wages is based on the optimal identification and definition of fair wage developments. With this in mind, 12 fair wage dimensions were identified (see Table 3.1 next page).

Capturing complementarities between wage issues

All the fair wage dimensions proposed here are needed in the sense that they help understand different parts of the company's wage story. If one dimension was missing, any auditing may risk not capturing the real nature of wage practices at the company. While all these indicators are complementary, they may not move in the same direction. For example, although a significant increase in real wages may help to improve the company's performance with regard to other wage indicators, such as the living wage, it does not ensure that there would not be a significant decline in the wage share or an increase in wage differentials. Similarly, the minimum wage may not mean much if we do not know what is offered in terms of non-monetary benefits, which may help workers to cope.

The minimum wage is also closely related to the living wage. While the minimum wage is essential for ensuring the living wage dimension, it may not be enough. Statistics show that, in many countries, the minimum wage remains well below the national subsistence minimum or poverty line (ILO 2008b). Therefore, even workers who earn the legal minimum wage may find it insufficient to meet basic needs.

Moreover, many workers' concerns extend well beyond mere 'bread and butter' issues and relate more to obtaining recognition from their employers in terms of their work and skills (through better pay systems), or better communication (work contracts, communication on wages) and establishing a permanent relationship with their employers, which would allow them to revisit and renegotiate wages and employment on a regular basis (social dialogue).

The picture of fair wages would also be incomplete without taking working hours into account. When workers are paid very low wages, they have a tendency to work excessive hours in order to make ends meet. Efforts to reduce overtime without tackling poor wages can, therefore, be problematic. On the other hand, efforts to increase wages are not enough without counting how many working hours are worked and whether overtime payments are properly remunerated. Similarly, the tendency to work excessive working hours is frequently due to the pay systems themselves, often based on the piece rates that induce workers to accumulate working time and minimise rest time in order to obtain the highest possible wages.

Table 3.1 The 12 fair wage dimensions

A fair wage could be defined as:	
1. Payment of wages	A wage that is regularly and formally paid in full to the workers.
2. Living wage	A wage that ensures minimum acceptable living standards.
3. Minimum wage	A wage that respects the minimum wage regulations.
4. Prevailing wage	A wage that is comparable to wages in similar enterprises in the same sector.
5. Payment of working time	A wage that does not generate excessive working hours and properly rewards normal working hours and overtime.
6. Pay systems	A wage that leads to a balanced wage structure/ composition between the basic wage and additional bonuses and benefits. A wage that reflects different levels of education, skills and professional experience, as well as rewarding individual and collective performance. A wage that complies with regulations on social insurance payments and paid holidays and is not dominated by disciplinary wage sanctions.
7. Communication and social dialogue	A wage on which workers receive sufficient information in advance (through an individual work contract), in the course of the production process (through regular communication channels) and at the time of the wage payment (with a detailed pay slip). A wage that is negotiated individually (with individual employers) and collectively – notably through collective bargaining – between the employer and the workers' representatives who are freely accepted in the company.
8. Wage discrimination/ wage disparity	An equal wage for equal work that does not lead to wage discrimination and does not generate unjustified, too high and too rapidly growing wage differentials within the company.
9. Real wages	A wage that progresses at least in proportion to price increases.
10. Wage share	A wage that progresses proportionally along with enterprise sales and profit growth and does not lead to a fall in the wage share in enterprise performance growth.
11. Wage costs	A wage whose progression does not lead to a dramatic reduction in wage costs within total production costs and as a percentage of employment.
12. Work intensity, technology and up-skilling	A wage that progresses along with changes in intensity at work, technological contents and the evolving skills and tasks of the labour force.

Wage problems along the supply chain: 2010 Fair Labor Association auditing results

In order to compile more information on wages it was decided, with the assistance of the FLA, to add an important number of new questions on wages into their annual auditing process. After having tested this process in 2008–2009, an extensive auditing was carried out in 2010. This exercise allowed better identification of all the different wage dimensions defined in the second section and their interaction with other areas of the world of work, such as working time, production performance and labour force composition.

The auditing was carried out by the FLA in autumn 2010 among 122 companies, in Bangladesh, China, India, Indonesia, Pakistan, the Philippines, Sri Lanka, Thailand, Turkey and Viet Nam (see Figure 3.1). Since most suppliers of brands (estimated at 80 per cent) are located in China, the largest number of companies were audited in various Chinese provinces (63 companies or 51 per cent of the sample), while interviewing companies in other Asian countries and in Turkey helped to build some comparative wage analysis.

Compliance with national legislation

We begin by identifying those areas in which enterprises did not comply with existing national legislation. It was checked whether companies

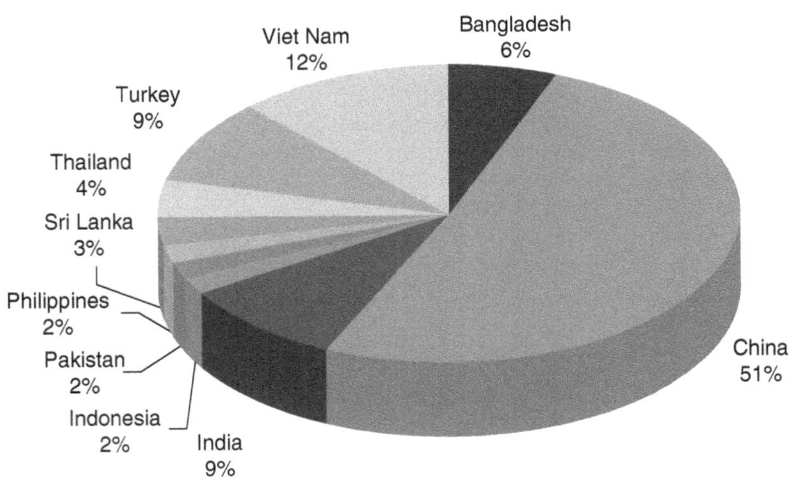

Figure 3.1 Regional distribution of suppliers

were producing fake records, involving dual or even triple record keeping. Forty-eight per cent of companies were found to be operating in this manner, with official records being prepared for the authorities and labour inspection (generally in compliance with all the relevant labour regulations), but with other sets of records for the brands and, sometimes, for other kinds of external auditing, such as FLA. The dual record process was found to be more prevalent in Bangladesh, China, Sri Lanka and Turkey (Figure 3.2).

Companies resorting to dual recording had compliance problems in terms of wages, working time and other working conditions. The same companies were generally also found to have a series of problems in other areas, for instance in terms of communication and information on wages, social dialogue, pay systems and payment of wages. Dual records on wages were often related to double records in working hours aimed at hiding excessive working hours beyond the legal limit and not being properly paid.

This was confirmed by questions on the payment of wages. While not paying wages at all was a problem reported in less than 2 per cent of companies, 17 per cent reported having been obliged to postpone wage payments. A total of 66 per cent also reported the underpayment of wages. This problem was particularly acute in India (82 per cent of companies) and in China (71 per cent), but was less common in Viet Nam.

A majority of these cases concerned the underpayment of overtime: 61 per cent of companies reported facing difficulties in paying overtime. This was a striking finding and confirmed the payment of working hours and overtime – as the main area of non-compliance with regard to wages. Because of fake records, reality may be expected to be even worse than the results reported in Figures 3.3 and 3.4.

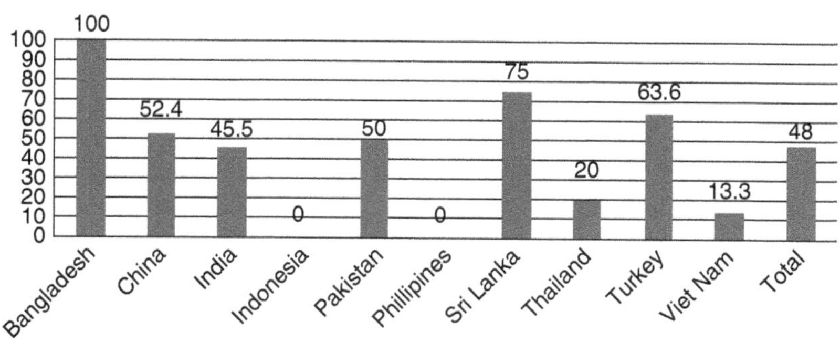

Figure 3.2 Enterprises with dual records (%)

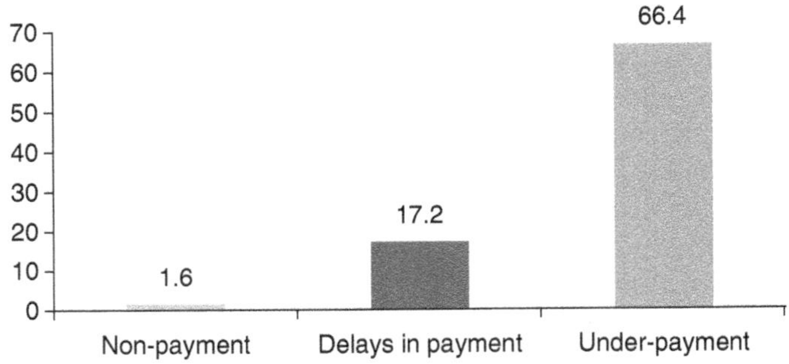

Figure 3.3 Enterprises with payment problems (%)

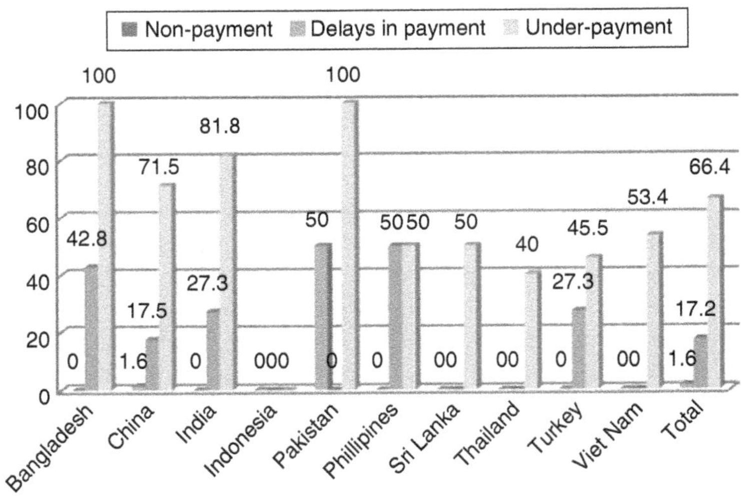

Figure 3.4 Enterprises with payment problems (%), by country

There were also problems of compliance with minimum wage regulations (Figures 3.5 and 3.6). Twenty per cent reported a starting wage that was below the official minimum wage, while a majority (75 per cent) proclaimed a starting wage of exactly the official minimum wage. Because of double record keeping on working hours it was difficult to check whether these companies were paying such an amount for the basic working hours as stipulated by law or including overtime. This could mean that they do not comply and pay a starting wage per hour below the official corresponding hourly minimum wage. Very

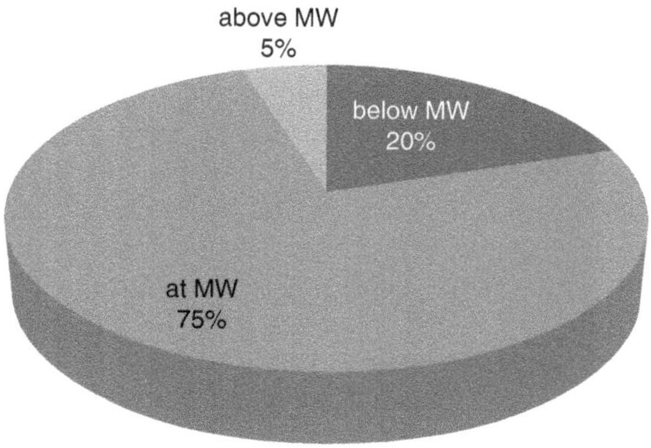

Figure 3.5 Companies' starting wage compared to minimum wage (MW)

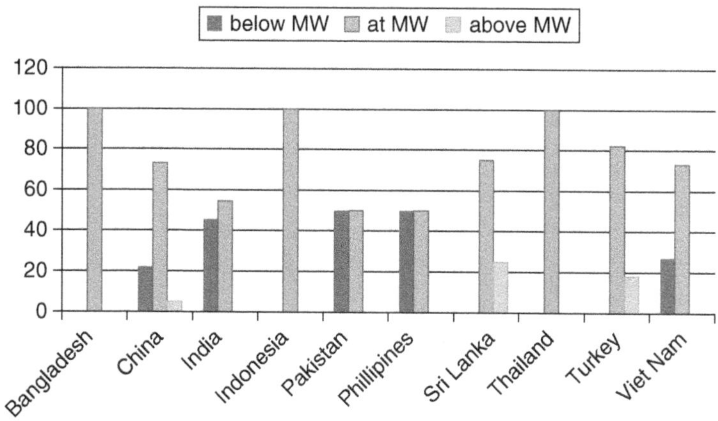

Figure 3.6 Starting wage compared to minimum wage (% of enterprises)

few enterprises (5 per cent) reported a starting wage above the official minimum wage. These results highlight the important role of the minimum wage since it seems to be directly influencing starting wage levels at enterprise level. The problem of non-compliance to the minimum wage seems to be particularly important in India (among 45 per cent of enterprises of this sample), but also in Viet Nam and China. On the other extreme, small percentages in countries like Turkey or Sri Lanka may also be due to less transparency and more difficulty in identifying the problem during the auditing.

The question of whether the enterprises were providing social insurance to all workers was also asked. A total of 43 per cent reported they were not, often providing insurance only for a fraction of the labour force or leaving it to voluntary contributions from the workers. Many companies were, thus, found not to be complying with obligatory social insurance contributions (Figure 3.7). The percentage was the lowest in China, with only 17 per cent of enterprises providing social security contribution – which can be explained at least partly by the drawbacks of the current social security system, notably the difficulties involved in transferring accumulated social insurance contributions from one province to another. Enterprises in India with full social security coverage were also rare.

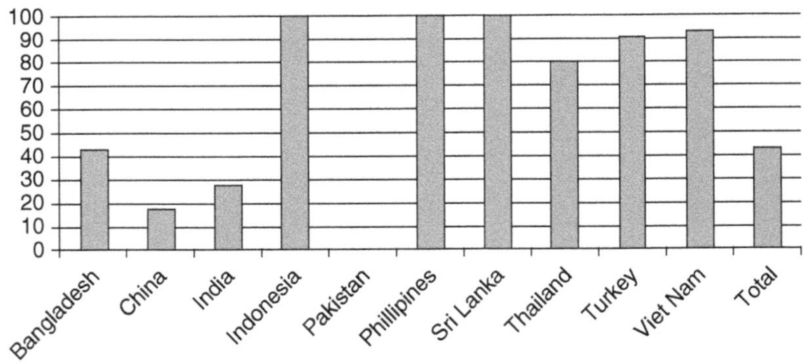

Figure 3.7 Enterprises paying social security contributions (%)

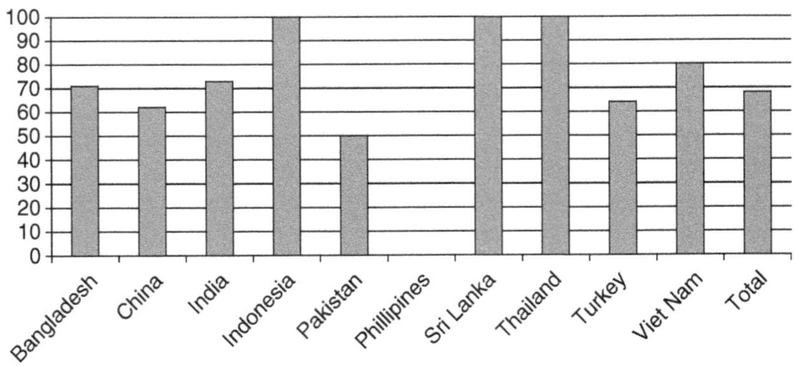

Figure 3.8 Enterprises providing paid holidays (%)

During auditing, we also investigated whether paid holidays were being granted in accordance with the law. Thirty-two per cent of companies were not providing such obligatory provision, this drawback being even stronger in Turkey, China and the Philippines (Figure 3.8). We should also emphasize that even where paid holidays are granted to the workers, the pay systems in place – for instance the piece rate – is pushing the workers to stay in the company to avoid losing a significant part of their income when they effectively take their holidays, during which they are paid only the minimum wage.

Living wages

Poor compliance with the minimum wage – which is expected to be fixed at a level allowing workers to satisfy basic needs – is very much related to the living wage dimension.

Questions asked of management relating to the living wages resulted in many uncertain answers. In particular, when asked the question 'Is the starting wage provided by your company sufficient to ensure workers a minimum living standard?', a majority of enterprises (75 per cent) reported that they did not know or did not answer at all. This confirms the difficulties involved in enterprises assessing and monitoring this wage function, as well as the need to develop a methodology to help management and workers to systematically monitor living wages. These findings did confirm problems in ensuring this basic wage function: 21 per cent of managers clearly stated that they believed the starting wage they were paying their workers was not sufficient to allow decent living standards, while only a very small proportion of companies – 4 per cent – reported that the starting wage they provided (generally corresponding to the official minimum wage) was adequate in this respect.

When we compared the starting wage at enterprise level to the Asia Floor Wage,[4] nearly all enterprises were paying not only a starting wage but also an average wage below the threshold defined by the Asia Floor Wage for their country. The exception was China where a higher percentage of enterprises had an average wage fixed above the Asia Floor Wage, an aspect that shall be investigated in more depth in the case studies section.

This low pay problem among suppliers was confirmed by answers to questions about the dimension on the prevailing wage (Figure 3.9). Interestingly, it was found that there were more companies paying below (45 per cent) than above (33 per cent) the prevailing wage. Only 22 per cent were paying the exact industry prevailing wage.

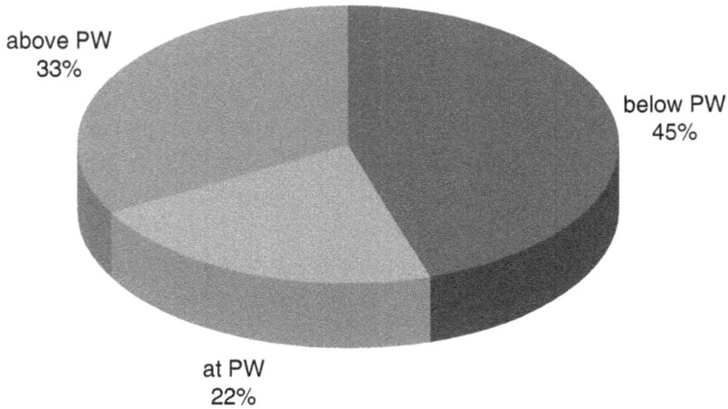

Figure 3.9 Payment of prevailing wage (PW)

Data on wages also helped to identify the wage gap between employees at the top (managers) and those at the bottom (unskilled production workers). The results (Table 3.2) show that enterprises in the Asian countries were found to have a very low disparity between the highest and the lowest wages, generally of less than six times – much less than in Western companies, where the wage gap can reach 30 times (ILO 2011). Higher wage disparity (1 to 10) was found among the few Indonesian companies, and among suppliers from Viet Nam (7.7). This low wage disparity was confirmed by the case studies presented. Moreover, no discrimination was found between men's and women's average wages in the audited companies, while some cases of discrimination against migrant workers were reported in a few companies in China and Indonesia. Interestingly the wage data collected over three years – from 2008 to 2010 – show that wage disparities did not increase – or even very slightly decreased – during this period, highlighting that the management strategy within the crisis was rather aimed at maintaining and even increasing wages of low skilled production workers – even if to the detriment of other categories – in order to keep the production process. However, wage disparity tends to increase over the last few years especially in China (Vaughan-Whitehead 2010).

Working time

While wage levels were found to be low, auditing also revealed that they were the final wages paid to the workers and, thus, also included overtime. Dual records found in half of the companies are an indication of

Table 3.2 Wage disparity between workers at the top and those at the bottom (between highest and lowest wages), 2010

Bangladesh	China	India	Indonesia	Pakistan	Philippines	Sri Lanka	Thailand	Turkey	Viet Nam	Average
5.9	5.9	2.5	10	n.a.	1.6	1.9	n.a.	n.a.	7.7	**5.7**

Source: Auditing on wage issues, FLA 2010.

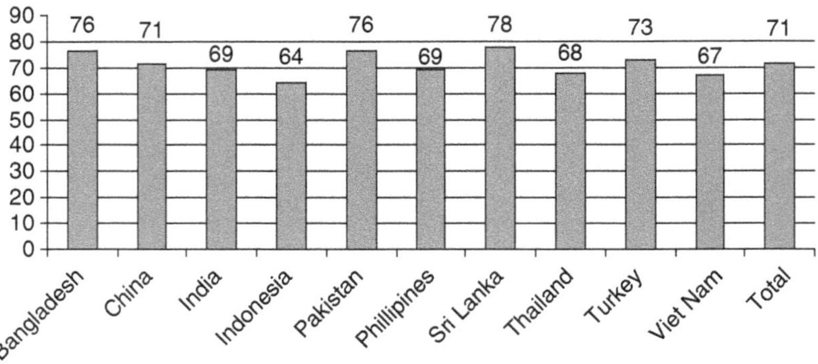

Figure 3.10 Number of hours worked per week

the difficulties of identifying the exact number of overtime hours and how much of those are paid. Nevertheless, the auditing did allow source measure of extended working time (Figure 3.10). First, in the majority of enterprises (88 per cent) workers were requested to work more than the legal 60 hours a week and to work more than six days in a row (66 per cent).

Average working hours were 71 hours a week for the whole sample, with an even higher number of weekly hours in Turkey (73), Pakistan (76), Bangladesh (76) and Sri Lanka (78). Workers in the case studies confirmed that they had to work overtime if they wanted to increase their wages to acceptable levels for ensuring basic living standards.

The problem is that overtime was also found to be massively underpaid, with a majority of enterprises (61 per cent) found not to pay overtime hours according to the existing legal provisions (Figure 3.11). This problem was most acute in China (65 per cent of enterprises), India (82 per cent) and Pakistan and Bangladesh (100 per cent but on a small sample that cannot be considered representative).

Wage adjustments

After questions about wage levels, we investigated wage evolution within the company in order to better identify what criteria were taken into account when adjusting wages from one year to the next: prices, profits/sales and labour and capital changes. Through auditing it was possible not only to check wage levels but also their adaptation to a number of indicators, such as price increases and enterprise performance. The findings confirmed the generally static nature of wages.

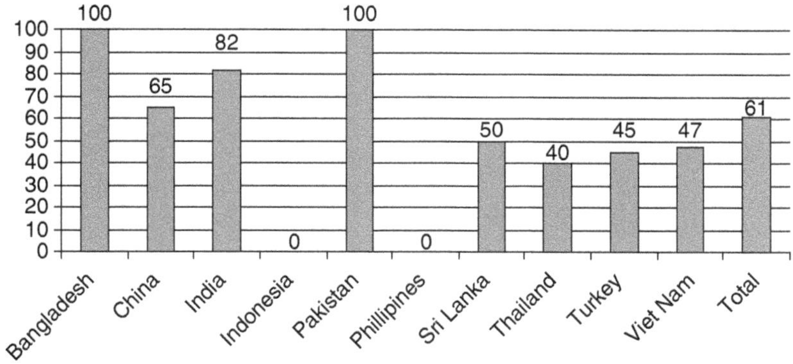

Figure 3.11 Percentage of companies underpaying overtime

Using real wages is essential since inflation pressures have risen in some Asian countries since 2009 (ILO 2011).

In real wage terms a significant percentage of companies were found to have increased nominal wages (91 per cent), at a general average rate of 16 per cent over the period 2008–2010 (Table 3.3). However, such increases remained below the rate of inflation in a number of countries (Sri Lanka, the Philippines, Viet Nam and Indonesia). In particular, wages in Sri Lanka and Viet Nam did not catch up with high inflation rates. By contrast, relatively low inflation rates in China and Thailand in 2009 led to nominal wages generally higher than the inflation rate.

In 2010–2011, rising prices especially of food items increased pressure on companies to take inflation into account when adjusting wages. Moreover, the fact that inflation seems to be mainly driven by commodity prices especially on foreign energy and food items – particularly in India, Indonesia, the Philippines and Thailand – and not at all due to wage increases shows that it would be possible to increase wages without further fuelling price increases (IMF 2010).

Very few enterprises reported to have linked wage increases to profits or other indicators of the company's performance. Furthermore, all data collected on wages and on enterprise performance have shown that there was a poor connection between the two; that is, wages have not been increased along with higher enterprise performance. At the same time, some differences could be observed between the pre-crisis and the crisis periods. While before the crisis wage growth was found not to have increased in step with profit growth, some downward wage flexibility was observed during the crisis, with a number of enterprises

Table 3.3 Nominal wages increases compared to price increases, 2008–2010

	Bangladesh	**China**	**India**	**Indonesia**	**Pakistan**	**Philippines**	**Sri Lanka**	**Thailand**	**Turkey**	**Viet Nam**
Wages 2008–2010	14.5	20	17.1	12	n.a.	4.6	18.9	n.a.	n.a.	25.2
Inflation 2008–2010	14.8	5.2	18.1	15.1	35.3	12.8	26.8	4.6	17.3	31.3

Source: Auditing on wage issues, FLA 2010.

that experienced some decrease in profits and sales cutting the wages of their workers. In contrast, no upward correlation was found in those companies that continued to report increases of profits in 2008–2009.

Poorly motivating pay systems

Beyond wage levels and adjustments the nature of the pay systems themselves was also investigated, and information was obtained on the different wage systems used by the companies (Table 3.4). The first, rather striking, result was the dominance of the piece rate system. Not only did a great number of companies (38 per cent) use a piece rate system (especially in China, 49 per cent of them), but a high percentage of workers' total wages are paid in this way: 80.5 per cent of total wage on average. Seventy-two per cent of enterprises with such a piece rate system distribute more than 80 per cent of total income in this way. This means that there is no basic wage that could, for instance, be calculated in accordance with some sort of pay grid, with wage levels reflecting different skills, education and experience. This was confirmed by answers to the questions on the existence and operation of a pay and classification grid system: only 40 per cent of companies reported using a grid system, this percentage being even lower in China (27 per cent) and absent in Thailand and Turkey.

The predominance of the piece rate system (Table 3.5) also meant that the pay systems were rather unbalanced since they did not leave any room for introducing alternatives.

The auditing exercise also revealed how poorly developed (in only 2 per cent) pay systems related to enterprise performance are in comparison with those based on individual performance, although such systems are used in some Asian countries, as well as Western ones, and have been shown to motivate workers and to boost productivity. The lack of a pay system related to enterprise performance may represent one cause of this disconnect between wages and productivity. Forty-two per cent of enterprises preferred to distribute 'production bonuses' or 'efficiency bonuses', but on a discretionary basis without using very clear criteria for measuring them. There were also some enterprises distributing bonuses related to seniority (23 per cent). Attendance bonuses were widely used, by 51 per cent of enterprises, generally in complement to the piece rate system (Table 3.6). The percentage of enterprises with both seniority bonuses and attendance bonuses was found to be much higher among suppliers from Viet Nam and Sri Lanka. The case studies in this section outline the specific and very strict conditions (from 26 to 30 days' attendance required) on which the payment of

Table 3.4 Pay systems by country (use of disciplinary cuts and wage grids), 2010

% of enterprises with:	**Bangladesh**	**China**	**India**	**Indonesia**	**Pakistan**	**Philip.**	**Sri Lanka**	**Thailand**	**Turkey**	**Viet Nam**	**Total**
Wage disciplinary cuts	14	20	0	0	100	0	0	0	9	7	**15**
A wage grid	86	27	55	50	100	0	75	0	0	93	**40**

Source: Auditing on wage issues, FLA 2010.

such attendance bonuses seems to depend, especially among Chinese suppliers.

At the same time, a majority of enterprises (75 per cent) were found to pay at least one form of non-monetary bonus (Table 3.7), which can be an important source of non-monetary income and significantly increase workers' living standards. This percentage was even higher among the audited companies from Sri Lanka, Thailand, Pakistan (100 per cent but on very small sub-samples), Viet Nam (94 per cent), Turkey (91 per cent),

Table 3.5 Piece rate (PR) systems, by country, 2010

% of enterprises with:	A piece rate	In % of their total wage	% where PR more than 80% total wage
Bangladesh	0	–	–
China	49	86	37
India	27	57	0
Indonesia	50	100	100
Pakistan	100	45	0
Philippines	0	–	–
Sri Lanka	0	–	–
Thailand	0	–	–
Turkey	27	32	0
Viet Nam	40	97	83
Total	**38**	**80.5**	**72**

Source: Auditing on wage issues, FLA 2010.

Table 3.6 Wage structure (bonuses), by country, 2010

% of enterprises with:	Profit-sharing	Efficiency bonus	Seniority bonus	Attendance bonus
Bangladesh	0	14	0	100
China	3	54	29	46
India	1	0	27	36
Indonesia	0	50	0	50
Pakistan	0	50	0	0
Philippines	0	50	0	0
Sri Lanka	25	25	50	100
Thailand	0	60	0	80
Turkey	0	27	0	0
Viet Nam	1	40	33	86
Total	**2**	**42**	**23**	**51**

Source: Auditing on wage issues, FLA 2010.

Table 3.7 Provision of non-monetary benefits, by country, 2010

	Bangladesh	China	India	Indonesia	Pakistan	Philippines	Sri Lanka	Thailand	Turkey	Viet Nam	Total
Average number of NM benefits provided	2.6	2.7	1	1.5	2	0	4	3.8	2.8	2.9	2.6
Full (100%) NMB potential	28	30	11	17	22	0	44	42	31	32	30

Bangladesh (86 per cent) and China (68 per cent). A higher percentage of enterprises in China, however, were providing free accommodation, where non-monetary benefits traditionally complement the monetary wage. Enterprises providing the higher number of fringe benefits were found in Sri Lanka (an average of four non-monetary benefits) followed by Thailand (3.8) and Viet Nam (2.9). This remains, however, below the total range of non-monetary benefits that could be provided by the companies including accommodation, food, medical care, health coverage to transportation and other minor advantages.

The largest number of companies (61 per cent) provided free meals. Fewer than expected provided accommodation (42 per cent) and even fewer transportation (10 per cent) and medical services (6 per cent). The percentage of enterprises providing accommodation was much higher in the sub-sample on China (among 57 per cent of audited companies there), which might be explained by the high worker mobility and the importance of migrant labour moving from one province to another. In most garment suppliers in China, migrant workers generally represented between 70 and 100 per cent of the total labour force.

Poor communication and poor dialogue

In order to assess wage fixing a series of questions was also asked about communication and social dialogue. First, in terms of communication, 45 per cent of enterprises reported having signed individual formal labour contracts with the workers. The majority of companies (93 per cent) were found to provide a pay slip to individual workers. A lower percentage indicated the total number of hours worked on the pay slip, however (48 per cent); or distinguished between regular working time and overtime (48 per cent). Even fewer (32 per cent) were found to detail individual performance according to different piece rates. To summarize, even when pay slips are distributed, they do not always – in at least 50 per cent of audited enterprises – provide the necessary information, with at least one key piece of information missing (Table 3.8).

This poor quality of information was confirmed by the high number of companies (52 per cent) where workers were not aware of how their wage and benefits were calculated (Table 3.9). The percentage of companies in this situation was particularly high in Bangladesh, Indonesia and Pakistan (100 per cent but on very small and, thus, not representative sub-samples), but also Sri Lanka (among 75 per cent of enterprises) and India (64 per cent).

In terms of social dialogue, a relatively high proportion of enterprises reported having a trade union (42 per cent) and fewer a workers'

Table 3.8 Communication on wages, by country, 2010

	Bangladesh	China	India	Indonesia	Pakistan	Philippines	Sri Lanka	Thailand	Turkey	Viet Nam	Total
Providing a pay slip (%)	86	95	64	100	100	100	100	100	100	100	93
Average number of pieces of information on pay slip (out of 3*)	3	2.5	1.1	2.5	1.5	3	2.2	3	2.5	2.9	2.4

*The three types of information checked in the pay slip were:
1. Information of total number of hours worked.
2. Distinction between basic working hours and overtime.
3. Indication of piece rates applied for calculating piece rate payments (where piece rate).

Table 3.9 Percentage of companies where workers were aware of their wage and benefits

Bangladesh	China	India	Indonesia	Pakistan	Philippines	Sri Lanka	Thailand	Turkey	Viet Nam	Total
0	48	36	0	0	100	25	20	91	73	48

Table 3.10 Social dialogue practices, by country, 2010

% of enterprises with a:	Trade union	Workers' committee	Collective agreement
Bangladesh	0	57	0
China	54	19	13
India	0	36	0
Indonesia	50	50	0
Pakistan	50	100	0
Philippines	50	50	50
Sri Lanka	0	75	0
Thailand	0	0	0
Turkey	0	82	0
Viet Nam	93	47	80
Total	42	35	17

collective (35 per cent) (Table 3.10). Only 17 per cent had a collective agreement. Interestingly, 52 per cent of companies reported that wages in their company were fixed unilaterally by the employer, meaning that no individual nor collective negotiations took place with the workers.

The auditing exercise clearly helped to identify the countries where there is generally a total lack of trade unions: India, Sri Lanka, Turkey – where workers' collective prevail instead – and to a lesser extent Bangladesh and Thailand (due to the too small sample to reach definite conclusions). By contrast 93 per cent of companies in Viet Nam, 54 per cent in China and 50 per cent in Pakistan and Indonesia were found to have a trade union.

However, qualitative case studies presented in the section below have highlighted the fact that these trade unions are rarely involved in wage issues, not to mention other working conditions. Moreover, few collective agreements have the aim of establishing clear conditions in wage fixing and other working conditions – although this is typically their primary aim.

Case studies: confirming the need for a multi-dimensional approach

The results presented in the previous section have shown that companies' wage practices and their fair wage performance vary by country. For example, in China individual companies were particularly reliant on the piece rate system, and were also found to be poor in terms of communication, information and social dialogue on wages. Different wage problems prevailed in other countries, such as lack of wage adjustments to growing inflation in Viet Nam and in Sri Lanka. The fair wage approach helps to better pinpoint the problems in different wage areas and then to identify possible remedial action. The same approach also highlights important variations between companies within one country, which we look at in the analysis in this section.

Methodology

Three companies were randomly selected out of the initial sample of enterprises in China considering the larger sub-sample of suppliers in this country within the total sample. The case studies were carried out in February–March 2011, with extensive interviews with both management and workers (split into skilled and unskilled workers' focus groups). An average of two days was spent in each company. In addition, each factory was asked to fill in a 'statistical wage table' to collect data on employment, wages and enterprise performance over the previous three years (2008, 2009 and 2010). It is on the basis of these three sources of information that assessment for each enterprise of its performance on each of the fair wage dimensions was possible.

The three companies are described in the section below to better identify their specific features, before providing a cross-enterprise assessment to identify both common and divergent practices in the wage area (Table 3.11).

Description of the three companies

Company A was established in 1989 to produce sports shoes. It is owned by a Taiwanese company, 25 per cent municipally-owned by the province and 10 per cent state-owned. Companies B and C produce knitted apparel. Company B is owned by a Malaysian group, which owns factories across South-East Asia and employs more than 13,000 people in the region. Company C is owned by a large Chinese garment group from Ningbo and

Table 3.11 Overview of the three companies

	Factory A	Factory B	Factory C
Number of employees	2978	332	2329
Main products	Shoes	Woven and Knitted apparel	Knitted apparel
Ownership	Mixed	Foreign	National
Location of parent company	Taiwan	Malaysia	China
Years in operation	22	4	12

employs 40,000 people across China. This Chinese group is mainly supplying a Japanese brand so it has been inspired by Japanese corporate culture.

The data confirmed that Chinese companies were hit by the economic crisis even if they succeeded to maintain a high rate of economic growth. Profits and sales slowed in 2008–2009. Considerable employment adjustments occurred, from 4 per cent in company A to 64 per cent in company B. Surprisingly, all these changes did not arise because of manager initiative – to layoff in order to adapt to lower sales during the crisis – but rather from a very high worker mobility because of labour shortages that appeared despite the economic slowdown, and which increased with economic recovery in 2010. Significant employment reduction was explained by management of company B as voluntary departures, and this was confirmed by workers themselves. This explained why nominal wages increased significantly from 28 to 48 per cent (in companies C and A respectively B being in the middle with 31 per cent increase), which also led to a substantial increase in real wages in a context of low inflation and even negative rates in 2009. This wage increase was also facilitated by a government decision to increase the minimum wage in 2010 after having kept it frozen in 2009. Nevertheless, a number of different wage problems were identified among the three suppliers (Table 3.12).

Table 3.12 Main employment and wage developments in the three companies, China

2008–2010	Factory A	Factory B	Factory C
Employment variation 2008–2010	–4%	–64%	–20%
Starting wage in 2010 (in RMB)	1500	1800	1100
Average wage in 2010 (in RMB)	2086	2100	1880
Wage variation 2008–2010	+48%	+31%	+28%
Lowest wage 2008–2010	+25%	+78%	0%
Highest wage 2008–2010	+13%	0%	8.6%
Wage gap top/bottom	Decreased: 4.8 times in 2008 4.4 times in 2010	Decreased: 11 times in 2008 6 times in 2010	Increased: 2.6 times in 2008 2.9 times in 2010
Wage costs variations 2008–2010	Increased	Increased	Increased

Common wage issues

The case studies confirmed a number of problematic wage issues earlier identified from the audit results. A number of issues were common to the three companies under study.

Low wages

While no problems were reported in the regular payment of wages, there were cases of suspected under-payment of overtime. Moreover, in one company there were disciplinary wage cuts if overtime was not accepted by workers. In terms of adjustments, wages were found to have increased in both nominal and real terms so that the companies performed rather well on the real wage and wage costs dimension. At the same time workers reported that their wage levels continued to be too low, and that they needed to work overtime to ensure decent living standards.

In all companies, workers indicated that the provision of non-monetary benefits helped them to cover their basic needs, thus, confirming the value of such elements in the companies' pay systems or packages. However, in all three companies, the monetary wage was not sufficient in itself to cover basic living standards, especially in basic areas such as health care and education.

Cases of non-compliance to minimum wage regulations

Case studies in China also confirmed a potential problem in respecting the legislation on the minimum wage that stipulates that workers should be informed about the statutory minimum wage prevailing in the province in which they work, and that such a minimum wage should be ensured for the normal number of working hours (i.e. 40 hours a week in the case of China, excluding overtime). On the contrary, not only were workers paid only slightly above the legal monthly minimum wage, but wages were paid against a total number of working hours much higher than 40 hours a week (evidence in companies A and C, and suspects but no evidence in company B).

Excessive overtime not always properly remunerated

The three companies were found to have excessive working hours. Workers in company A not only reported 32 hours of overtime in the week, but also that they were not paid according to the legal regulations. In particular, work on Sunday was not always paid as overtime. Moreover, at the end of their eight normal hours they were told how many hours of overtime they were expected to do and they had no choice but to stay if they did not want to experience disciplinary wage cuts.

A wage fixing system dominated by management

The wage fixing system was found in some cases to be unbalanced due to an over-reliance on overtime (especially in company C), but also because of an over-reliance on the piece rate system for the two companies that mainly used it. In one company the piece rate represented 100 per cent of workers' total wage. The attendance bonus also required a high number of days at work (26 days). Moreover, wages were not found to be somehow related to skills or to enterprise performance. In two out of three companies there was no wage grid. This was particularly problematic in company B, where management identified worker recruitment and retention as its main priority. As workers are learning new skills or as the company is performing well, workers' value increases given their new skills or increased contribution to company performance and as such should be reflected in their wages. Greater intensity at work has also not been leading to wage increases. In company A for instance, the production has doubled from 100 shoes to 200 shoes an hour, partly due to new technology but also to higher intensity at work.

In one company wage discrimination was reported between temporary and permanent workers, and in another an increase of wage disparity between workers at the top and those at the bottom was also highlighted by workers. In two companies, employees complained about differences between workers paid on a basic wage (hourly base) and those paid exclusively on the piece rates; first because the coexistence of the two systems created confusion and second because piece rate workers often were paid more than those paid on a fixed basic wage.

Lack of social dialogue

Finally, in all three factories – either explicitly emphasized by workers or implicitly observed by interviewers – there was a lack of social dialogue. In two out of three companies the workers reported that – despite the presence of a trade union – wages were unilaterally fixed by the management without any negotiation process. Moreover, in all three companies workers were unaware of their pay system and of their different bonuses; thus, reflecting a lack of internal communication on wage issues.

Differences in fair wage performance

Beyond these common features, wage problems also appeared to be rather different between the three companies, according to their own specific context and also the pay and human resource system each had in place.

Company A: low pay, living wage problems and excessive compulsory overtime

While the problem of too low wages was reported in the three companies, it was particularly acute in company A, where the starting wage was in fact the lowest and the average wage also slightly below the prevailing wage at industry level. Workers in company A directly complained of excessively low wages. Workers in particular expressed difficulties in covering their basic needs, especially for education and clothing. Affording medical care was also an issue. Indeed, while workers' physical injuries could be cured in the company's medical centre, basic medical conditions such as a cold or a sore throat were not eligible for free treatment and, at a cost of RMB 200–300, became prohibitive. Workers also stated that when they needed an important medical treatment they had to go back to their home town where medical care was cheaper.

These low wages and high levels of worker dissatisfaction may explain why the management had increased wages by 48 per cent over the previous two years. Despite this, the average wage in the company remained well below the prevailing wage and workers continued to complain about their low wages.

Workers in company A also complained about too many overtime hours – 32 hours a week. Moreover, they did not know before the end of their normal working day how many extra hours they would have to stay for. This was usually more than 60 hours a week and generally more than six days in a row. It was also found that overtime was not paid according to legal rates.

In addition, workers complained about the pay system. First, because they found it strange that some categories of workers were paid according to the piece rate system while others were paid on an hourly basis. Second, their wages were not related to their education and skills, with a lack of wage grid in the company also confirmed by management. The presence of a trade union did not impede the management to unilaterally define the wage fixing mechanisms and wage adjustments. Cases of wage discrimination, not only between those on a piece rate and those on a fixed basis but also between temporary and permanent workers, were also reported by a number of workers.

This combination of wages and working conditions led many workers to express their wish to leave the factory, something they had not done yet because most of their relatives were also working in the same province. One employee clearly stated that he was waiting for a factory to be built in his home province. In such a context, the company may face serious labour shortages in the near future.

Company B: high turnover despite improved wage levels and pay systems

In a context of very high turnover rates – with a 64 per cent decline in employment level in two years – the management of company B had implemented a number of measures to retain workers by improving both the level of wages and pay systems. Wages increased by 31 per cent and unit wage costs by 80 per cent between 2008 and 2010. Most wage increases targeted unskilled workers, and the lowest wage in the company was fixed at a higher rate than the province's minimum wage. As a result, in 2011 wages in company B had become the highest among the three companies under study. Non-monetary benefits were also provided, including free accommodation, free meals and medical care.

The management also reformed the pay systems, notably by diversifying the structure of wages, with a basic wage determined along a classification and wage grid being complemented by a series of bonuses, such as for seniority and attendance, and a collective scheme related to profits. Additional working hours – for which the workers reported to be properly paid – helped to increase the basic wage. Unfortunately these reforms were not accompanied by sufficient communication and information on wages, so that workers admitted not being aware of how their wage and benefits were calculated. They could not answer most of the wage questions. Employees continued to have the feeling that their wage was not reflective of their different skills and professional experience, and they were not aware of the collective bonus, for instance. The trade union does not seem to be involved in wage issues, so that workers reported that wages continued to be unilaterally decided by the management. Workers also complained about more complex tasks and higher intensity at work – due to more frequent requests by the brands for changes in the style of goods produced – that were not accompanied by wage increases, and that may explain the high rate of voluntary departures in the company.

Company C: piece rate system leading to long working hours, stress and under-payments

In company C, workers were exclusively paid under a piece rate system, which represented 100 per cent of their total wage. This over-dominance of the piece rate system creates income uncertainty for workers who tend, as a result, to work excessive hours on high work rhythms. Despite such a large number of working hours the average wage remained the lowest among the three companies. The starting wage was found to be exactly the legal minimum wage, but it included overtime; thus constituting a clear case of non-compliance to the minimum wage regulations, which stipulate that the minimum wage should be earned by workers for basic working hours excluding overtime. Nevertheless the employees reported having enough income to live on thanks to all the non-monetary benefits provided by the company, including free accommodation and medical services at the workplace.

The fact that all payments were based on piece rate meant that the wage structure was not diversified, with no additional bonuses and no wage grid. Employees, especially the most skilled, complained about the failure to link wages to education and skills levels, which they found particularly de-motivating. Also, communication on wages remained very poor, with an absence of collective bargaining and even discussions over wage issues in the company.

Different wage levels across companies

The wage levels were different among the three companies. Not only were their starting wages were different, from RMB 1,100 in company C to 1,800 in company B, but their average wages were also different, the highest being again in company B (2,100) and the lowest in company C (1,880).

These levels were compared to the Asia Floor Wage (AFW) threshold, which was calculated as being RMB 1,639 in 2008-2010 before being increased early in 2011 to RMB 1,842. Interestingly, the starting wage in each of the three companies was found to be below that threshold. At the same time, however, the average wage in each of the three companies was found to be above the AFW threshold, reflecting the high wage increases over the last few years. Labour shortages pushed enterprises to concede significant wage increases. Interestingly, wages of unskilled workers were increased much more than those of other categories, as shown in Table 3.13. Top-down wage disparity decreased in companies A and B and slightly increased in company C where the initial gap was already very low (2.6 times) especially compared to company B (11 times in 2008) (Table 3.14).

Pay systems: from diversified (company B) to single wage structure (company C)

Significant differences were also observed in the wage structure of the three companies, generally reflecting different pay systems.

Table 3.15 describes the wage structure prevalent in each factory. Pay systems varied, with some paying workers by the hour and others by the piece. These case studies, thus, presented a diversity of pay systems: company C exclusively used piece rates with all workers' income being

Table 3.13 Companies' starting and average wages compared to the Asia Floor Wage, China

In RMB	Company A	Company B	Company C
Starting wage 2010	1 500	1 800	1 100
Average wage 2010	2 086	2 100	1 880
Starting wage/AFW	91%	110%	67%
Average wage/AFW	127%	128%	115%
Asian Floor wage for China 2008–10: 1,639 (1,842 for 2011)	Starting wage below and average wage above the AFW	Both starting wage and average wage above the AFW	Starting wage well below and average wage above the AFW

Table 3.14 Top-down wage gap in the three companies

In RMB	Company A	Company B	Company C
Wage gap in 2008	4.8 times	11 times	2.6 times
Wage gap in 2010	4.4 times	6 times	2.9 times
Wage disparity in 2008–2010	Slightly decreased	Significantly decreased	Slightly increased

Table 3.15 Companies' wage structure, China, 2011

	Company A	Company B	Company C
Average wage	RMB 2,086	RMB 2,100	RMB 1,880
Basic wage	Yes for 54%	Yes	No
Salary grid	No	Yes	No
Piece rate	Yes (46% of workers)	No	Yes (100% of workers' total wage)
Attendance bonus	Yes	Yes	No
Collective bonus	No	Yes	No
Seniority bonus	No	Yes	No
Skill bonus	Yes	No	No

made variable and dependent on how many pieces were produced; company B paid its workers by the hour; and company A paid almost half of the labour force according to pieces and the other half by the hour.

Bonuses available to workers also greatly varied across factories, with company B having the most extensive range of bonuses and company C using no bonuses at all to pay workers. Finally, with regard to non-monetary benefits, free meals, access to subsidized accommodation and free access to medical care were very much the norm across the three factories in China. Those non-monetary benefits were found to represent an essential means of ensuring decent living standards for the workers and their families.

Differences reflected in final Fair Wage scoring

This study of wage practices in the three companies has allowed us to analyse their performance in the 12 fair wage dimensions initially defined. It has also helped to identify in each company the wage areas that are satisfactory and those that are problematic and where urgent action would be needed.

Table 3.16 presents such different fair wage assessments. The payment of wages (payment in time, regular payment, no wage arrears, no under-payment) was not much of a problem except in company A where serious under-payment of working hours was found. Questions put to workers about their capacity to ensure their basic needs in a number of areas (accommodation, food, education, health, etc.) complemented by a comparison of their starting wage and their average wage to the AFW has allowed an assessment of wages' living function.

Table 3.16 Fair wage performance of the three companies, China

Wage dimensions	Company A	Company B	Company C
1. Payment of wages	Needs improvement – unfair	Fair	Fair
2. Living wage	Needs improvement – unfair	Rather fair	Needs improvement – unfair
3. Minimum wage	Needs improvement – unfair	Rather fair	Needs improvement – unfair
4. Prevailing wage	Needs improvement – unfair	Fair	Needs improvement – unfair
5. Pay systems and wage structures	Needs improvement – unfair	Rather fair	Needs improvement – unfair
6. Social dialogue and communication	Needs improvement – unfair	Needs improvement – unfair	Needs improvement – unfair
7. Payment of working hours	Needs improvement – unfair	Rather fair	Needs improvement – unfair
8. Wage disparity	Needs improvement – unfair	Fair	Fair
9. Real wages	Fair	Fair	Fair
10. Wage share	Needs improvement – unfair	Needs improvement – unfair	Needs improvement – unfair
11. Wage costs	Fair	Fair	Fair
12. Wages, work intensity, technical and human capital	Needs improvement – unfair	Needs improvement – unfair	Needs improvement – unfair

Needs improvement – unfair

Rather fair

Fair

Note: Table prepared on the basis of the fair wage approach and its different fair wage dimensions and indicators (see the first section); for more details see Vaughan-Whitehead 2010.

It was clearly below the minimum standards in companies A and C although workers' reports were more encouraging in company C because of a wide number of non-monetary benefits. Higher wage levels and non-monetary benefits allowed workers in company B to ensure basic living standards.

The case studies confirmed a general problem of compliance to the minimum wage, since such levels were reached with overtime being included. This was clearly the case in companies A and C, with suspicion but no evidence of a similar process being found in company B.

Average wages were also below the prevailing wage in companies A and C and above in company B. Wage disparity between the top and the bottom of the wage scale did not seem to be a problem in the three companies. Cases of wage discrimination were identified though in company A between temporary and permanent workers, and not much transparency was found between wages of workers on a piece rate system and those under a fixed basic wage.

Company B was found to have attempted to introduce a better and more diversified wage structure, whilst pay systems in the other two companies were found to be demotivating and poorly efficient: in company A because of an excessive reliance on compulsory overtime and a dual wage system that was confusing for the workers; in company C because of an overreliance on piece rates that led to excessive stress and too many working hours.

In terms of wage adjustments, wages were found to have increased much more than price increases in the three companies so that real wages significantly increased, often to compensate for very low wage levels in 2008. This wage increase was reflected in the wage costs indicators, increasing in percentage of total production costs as well as in percentage of employment. In none of the three companies was a link with profits and enterprise performance found despite the attempt of company B to introduce a collective profit-related scheme, of which workers were, however, not aware.

Conclusions

Findings from auditing conducted by the FLA in 2010 presented in this chapter have painted a worrying picture. The statistics speak for themselves. Half of the enterprises in the sample continue to use dual records for wages and working time, and a significant percentage does not even comply with basic legal regulations such as the respect of minimum wage, payment of overtime, provision of paid holidays and social security contributions. Many continue to use illegal wage deductions. Non-compliance with laws on wages, thus, continues.

The auditing results have also shown a general problem of low wages, confirming a living wage problem that pushes workers to work longer hours to improve their take home pay. The rudimentary nature of pay systems, generally based on piece rate payments, and strict attendance bonuses further induce workers to work longer and lead to very heavy working rhythms and high intensity at work. The lack of social dialogue and communication on wages also help to explain the poorly motivating nature of the wage fixing system.

This assessment clearly stresses the need to finally address wage issues along the supply chain. It also shows the need to do so by capturing all the different pieces of the wage fixing process at company level: wage levels, wage adjustments, pay systems, wage bargaining and communication on wages, human resources in general. The complexity of the wage fixing process requires these issues to be addressed from a multi-dimensional perspective. The fair wage approach that has been used here has provided the right tool to capture the wage story within each enterprise, and for identifying the areas where progress is most urgently required and those where structural adjustments may also be needed.

The three case studies in this chapter, carried out in China, have confirmed the difficulty of reforming one wage dimension in isolation from the others. Although the three companies under study have significantly increased wage levels, mainly to deal with labour shortages, this was clearly not enough. In companies A and C pay systems continued to be characterized by excessive and under-paid overtime and be dominated by the piece rate system. In company B efforts to increase wages and to reform pay systems by introducing a profit-sharing scheme were poorly effective because management had not sufficiently involved workers and did not inform them about the advantages to be expected from the new pay system. As a result, the three companies under study continued to suffer from increasing voluntary departures, a problem that they will have to tackle not only by increasing wages but also by introducing reforms in the way wages are fixed, adjusted and communicated. Undoubtedly social dialogue and communication represent a cross-cutting area that will have to be developed in a more systematic way in order to make progress in all other fair wage dimensions.

The results presented in this chapter not only provide an overview of wage practices in the supply chain, but they also highlight some analytical tools and policy elements to start improving the situation.

The large scale and multi-dimensional nature of the problems identified here mean that issues will not be resolved without joint efforts and combined synergies between the different actors, actors whose respective responsibility on wages in the supply chain should be more clearly defined. Responsibility on the part of suppliers should be improved so that they take wage issues into account as a more central part of their planning and human resources processes. The brands must also be more responsive to including the complexities of the wage issues in their purchase orders; their demands for price, time, delays, pattern changes and conditions of delivery have enormous impact on suppliers' practices. Signals from brands of the need to improve wage practices is vital in pushing suppliers to move in the right direction.

The action of NGOs is also essential to ensure that fair wage practices are progressively introduced along the supply chain. The recent focuses given by NGOs to certain key wage issues – such as the living wage and the fair wage – have led to increased recognition of the need to do more in this area. The ILO, notably through its Better Work programme, also has an essential role to play. It can develop the necessary framework that will encourage better wage standards among suppliers, securing and amplifying such possible improvements through a general platform, on which all those concerned – brands and suppliers, NGOs, workers' representatives, academics, local and national governments – cooperate, ensuring that their actions in this field will, as far as possible, converge and be self-sustaining.

Concrete improvements are, thus, called for on the wage front, and not only enterprises – the suppliers and brands – but all CSR actors have to make it happen.

Notes

1. This analysis was done on the basis of data collected through the 2010 auditing process of the Fair Labor Association (FLA). I would like to thank Kenan Ercel, Patrick Kigongo, Jorge Perez-Lopez and Auret van Heerden for their kind cooperation. All the views expressed are those of the author and do not reflect the views of the ILO or of the FLA.
2. See for instance 'High street retailers slated for fashion's poverty pay', *Just. Style*, 8 October 2009; 'Ugly low-pay truth of high street fashion', *The Sunday Times*, 31 January 2010; 'Nike agrees USD 1 million overtime payment for Indonesian workers', *BBC News Business*, 12 January 2012; 'Cambodian workers hold "people's tribunal" to look at factory conditions', *The Guardian*, 2 February 2012.
3. See reports such as 'Cashing in – giant retailers, purchasing practices, and working conditions in the garment industry', CleanClothesCampaign,

February 2009; or 'Labour Behind the Label Report on wage payment', LabourBehindtheLabel, 7 October 2009.

4. The Asia Foor Wage is a campaign to set a concrete minimum living wage across Asia. Launched on 9 October 2009, the campaign calls on companies sourcing garment products in Asia to implement an Asian Floor Wage to improve the current wages of garment factory workers that do not cover basic needs and fall far short of a living wage. Although the Asia Floor is different in each country, it has the objective of covering the same set of goods and services in all countries. To see how the Asia Floor Wage is developing and is also stimulating the debate on wages along the supply chain, see http://asiafloorwage.org and also http://fair-wage.com/en/fair-wage-observatory/csr-on-wages/49-the-asia-floor-wage-campaign-developing-.html.

4

Regulating the 'Wage Effort Bargain' in Outsourced Apparel Production: Towards a Model[1]

Doug Miller

Introduction

In recent years efforts to address the vexed issue of wages in global apparel supply chains have been characterized by some interesting conceptual shifts and realizations. In acknowledging that on wages 'the needle has not moved very much' (FLA 2010: 4), the Board of the Fair Labor Association (FLA) recently endorsed an amendment to the compensation standard in their Code bringing their member companies (some of which also participate in the Better Work programme) nearer to committing to the notion of a 'living wage' as found in other multi-stakeholder initiatives, notably the Ethical Trading Initiative and Social Accountability International.[2] Significantly, this has been accompanied by the endorsement of a redefinition of the overall 'wages issue' into a more dynamic, multi-dimensional 'fair wages' concept in which questions of disparity, equality and broader social protection must also be considered (Vaughan-Whitehead 2010; Chapter 3 in this volume). This more sophisticated understanding and concomitant auditing methodology, which the notion of 'fair wages' brings to our understanding, and its potential for the regulation of wages in the apparel supply chain have to be welcomed. However, from what we know about the reality of outsourced apparel manufacture, three dimensions of wages continue to remain axiomatic in the context of the current crisis: the failure to pay prevailing wages on time and in full; the absence in many countries of a basic wage sufficient to meet basic needs and allow for discretionary expenditure; and the persistence of unrealistic production targets revealed in non-compliances in relation to excessive and often unpaid overtime and verbal and physical abuse of workers (Vaughan-Whitehead 2010; Bobo 2008; Bernhardt et al. 2008; Worker Rights

Consortium 2010; Jenkins 2010; Better Work Haiti 2010: 11, 2011: 16–18; Solidarity Center 2011; Better Work Jordan 2010, 2011; Better Factories Cambodia 2011[3]).

Running parallel to a decade long decline in apparel import prices in the United States and Europe (Francois et al. 2007; Global Macro Monitor 2011), company level corporate social responsibility (CSR) interventions and efforts by the Better Work programme to address non-compliances have tended to point to failures in both human resource and production management at the factory level and/or to weaknesses in national labour administration. Whilst cost-conscious buyers are particularly sensitive to minimum wage increases, in countries such as Bangladesh, where poverty wages remain an embarrassment for sourcing companies, a number of European and US buyers have sought, as in early 2010, to exhort the government to increase the minimum wage for the apparel sector. Similarly, efforts to address the problem of excessive unpaid overtime have generally focused on the lack of managerial capacity (Locke et al. 2009) and/or inefficiencies at factory level, and significant resources have been ploughed into assessing and, in some cases, improving factory productivity (GTZ, Deuster et al. 2009; Nathan Associates 2009).

These, of course, are all understandable and necessary interventions, but together they constitute an externalization of the problem of wage compliance and a failure to acknowledge the true economics of buyer–supplier relationships. As Nova (2013) has argued:

> The apparel industry is wedded to the existing production model and eager to deflect blame; it cannot acknowledge that its own pricing practices are the primary obstacle to progress.

This is an argument that has been echoed by a number of campaign groups specifically as an important first step towards the implementation of a 'living wage' in the garment industry (Labour Behind the Label 2009: 5; the Asia Floor Wage Campaign/Merk 2009: 60; Playfair Campaign 2008: 30–34). The Programme of Action submitted to the 10th World Congress of the International Textile Garment and Leather Workers' Federation[4] (ITGLWF 2009) called for 'responsible purchasing practices on the part of buyers based on long term business relationships, providing *sustainable pricing*'. In one of his final statements, the late Neil Kearney, general secretary of the global union, articulated this demand even further:

> A sustainable system would see the employer being responsible for the payment of a living wage and the buyer being responsible

> for making the payment of a living wage a contractual obligation, paying prices that enable the supplier to fulfill that obligation, and supporting suppliers in bearing the risk of paying higher wages for instance by providing greater stability in orders. This should not be an issue given the fact that wages make up such a small fraction of retail prices.[5]

Whilst a number of sourcing companies have begun to examine their purchasing practices, facilitated by their membership of multi-stakeholder initiatives, notably the Ethical Trading Initiative (2007 and 2010), the FLA (2008) and the former MFA forum, pricing policy, and more specifically labour costing, has until recently remained a neglected area. Moreover, collaborative initiatives in this area have been stifled by a mix of lack of political will, a lack of mode of implementation and perceived legal constraints, notably in the area of anti-trust and competition law.[6] However, the fundamental disconnect in efforts to achieve wage compliance, raise wages and address excessive overtime is that although supply factories, as employers, have prime responsibility for the compensation element of the employment relationship, their ability to uphold and progressively develop a compliant remuneration policy is heavily circumscribed by the multitude of ongoing negotiations about order price, payment terms, volume and lead time with, in some cases, a changing client base (cf. Impactt/Traidcraft 2009; Dirnbach 2009; Ruwanpura and Wrigley 2010). For buyers and their critics, the stark reality is that in a globally outsourced, competitive multi-buyer made to order production system, the nexus of wage compliance lies in the *individual transaction of an apparel order*. This means that unless a buyer wholly owns a manufacturing facility (Klein 2010), their ability to deliver on wage compliance, including a 'living wage', will always be compromised by the extent of their stake in overall production and by the duration of the existing and planned commercial relationship with their supplier factory. Consequently, for the majority of buyers, and for the subset that take social compliance seriously, the best they can do is factor in as accurately as possible the wage compliance costs in the manufacturing cost or CMT (cut, make and trim) price on any order they negotiate with suppliers. Figure 4.1 attempts to graphically represent this situation. In the supplier factory, wages (as is the case in many supplier countries) are determined by a government set minimum basic rate for the sector, which in many cases falls far short of national estimates of a living wage. Where a supplier holds to the 'artificially' set minimum wage, irrespective of its ability to pay more, in the absence

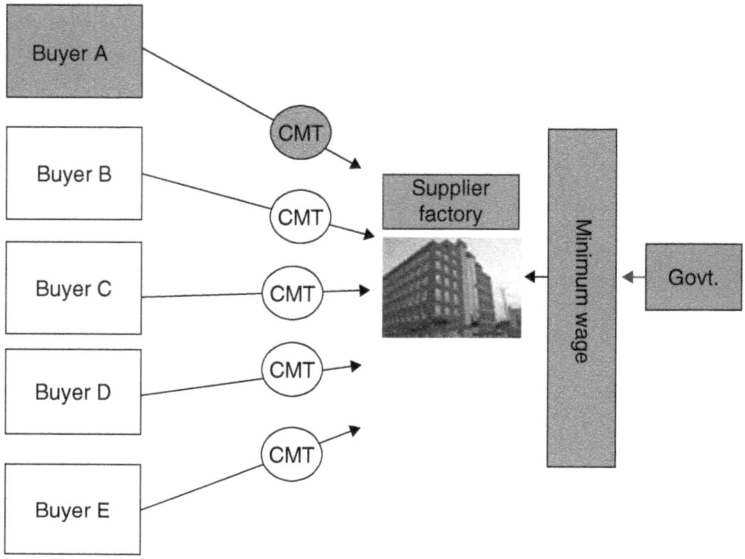

Figure 4.1 Why buyers cannot deliver a 'living wage'

of collective bargaining, the worker is unable to redistribute the profit generated by the supplier (and ultimately the buyer). The only remaining power is that of the local labour market, which results in high levels of wasteful labour turnover as workers move from factory to factory in search of better terms and conditions on offer. Buyer rents remain unassailable in a buyer-driven value chain, but for those sourcing companies committed to positively addressing the wage issue, their ability is circumscribed by both the size of their order in relation to other buyers in the factory and the duration of the commercial relationship. Thus, any move to address the wage effort bargain and to consider a mode of implementation for the delivery of a *commitment* to a living wage requires an examination of the determination of the CMT cost.

This chapter presents some preliminary findings on current labour costing practices; and, using an existing costing methodology, develops a model by which buyers might take appropriate action to progressively realize a level of compensation that meets workers' basic needs and provides some discretionary income. Since such a model raises a number of questions for buyers, suppliers, workers and trade unions as their representative organizations, these are discussed in a final section.

Costing garments and costing labour

Cost can be described as 'the economic value placed upon the economic resources consumed to make a product' (Jeffrey and Evans 2011). Historically, labour cost in garment manufacture has been determined by multiplying the estimated amount of time it takes an individual or group of workers to complete a garment by the actual labour cost – either the minimum wage or prevailing wage, or the actual earnings of a worker expressed as a minute value. This would normally be calculated by dividing the available wage figure by the number of minutes available on a daily, weekly or monthly basis. Historically in garment manufacture, workers have generally been remunerated by the 'piece' or on the basis of 'time achieved'. However, the estimated amount of time taken to complete a garment has always been a 'contested terrain' in the industry (Gomberg 1948; Edwards 1979; Edwards and Scullion 1982; Boggis 2001) going right to the heart of the twin compliance issues of wages and working hours for workers, and to the issues of effort, capacity, efficiency and ultimately profit for management.

In the hey-day of branded manufacturing, companies developed their own industrial engineering expertise to calculate labour cost and engage with workers in the so-called 'effort' bargain (cf. Behrend 1957). Three approaches emerged to determine a 'standard time' for assembling a garment: bespoke time and motions study, historical estimates and what is known as pre-determined time standards (PTS).[7] Bespoke time and motions study refers to the application of a range of 'techniques designed to establish the time for a qualified worker to carry out a task at a defined rate of working' (Kanawaty 1992: 243) that can only be carried out in-factory, with new styles made in the sampling department by an experienced machinist and the standard time and production target eventually established once assembly has actually commenced. The second method – PTS – is based on methods time measurement, whereby basic human motions are timed and used to build up a synthetic minute value for a given job at a defined level of performance (Kanawaty 1992: 381) under defined conditions. This approach deconstructs a garment into its constituent parts, identifies the manual labour operations required to complete these components and uses methods analysis to predetermine manufacturing standard times and production targets based on a systematic analysis of the operations required. So-called standard allowed minutes or SAMs are the time value targets set for a specific garment. These are based on an extensive sample of processes, physical environment, equipment and technology, and workers' remuneration observed in a range of manufacturing organizations. Often this is used interchangeably with the term standard minute value (SMV). Many manufacturers have invested in this approach to avoid time and motions studies. A third approach, preferred by those companies unable or unwilling to invest in industrial engineering, has tended to focus on approximations of labour time based on historical data.

Buying practice in costing labour

For global buyers, the traditional approach has been to consider the costs that make up the freight on board (FOB) or ex-factory price: fabric, trim, packaging and the manufacturing cost often referred to as CM (cut make) or CMT – cut, make and trim. CMT is a term used in the industry to describe a type of factory where the cost to the buyer relates to direct and indirect labour assembly, factory overhead and the factory profit. Where the focus is on quality of fabric from a particular supplier, the material will be already proscribed in the commercial contract, with the sourcing company focusing on the CMT element in the price negotiation. This breakdown of FOB into fabric and trim, plus CMT came about when garment importers moved from 'buying' to 'sourcing' under the MFA quota system (Birnbaum 2008: 17). Many buyers have key global accounts with fabric and thread manufacturers where the price is pre-negotiated. While buyers are inevitably drawn to consider the trimming of costs in every area, the CMT cost element is often the one element where there is room for negotiation, which can have a critical bearing on compliance.

There is no authoritative study as yet on buyer practice worldwide, but it would appear that as new brands, retailers and supermarkets entered the global apparel market in search of manufacturing capacity, they did so with underdeveloped knowledge of production management and industrial engineering principles. Preliminary evidence from a small scale survey and interviews with a number of practitioners in the United Kingdom and the United States indicate buying practices that do not separately itemize the labour cost in CMT (Miller 2010). To quote one manager:

> Most companies negotiate using historic data....Example you made that shirt for US$2.00 – make this one for US$1.90. Very little science goes into the negotiation and certainly 90 per cent of companies that work this way will not give a toss on (*care – author's translation*) what the labour rates are in the factory, as long as the external audits do not put them under the country laws of paying the 'minimum wage'.[8]

There is evidence of buyers insisting that their suppliers 'open their books' during price negotiations, a practice that some observers see as naked power play in an attempt to drive prices down. (Lamming et al. 2005: 558). In such circumstances, it is argued, suppliers may have no other option but to 'hedge' by quoting a CMT minute value that incorporates the overhead cost in with labour. Consequently, CMT negotiations can be a crude 'cat and mouse' exercise, (Lamming et al. 2005; SOMO 2003) as buyers continue to drive the market (Gereffi 1999) and their purchasing practices are based on target margins.

Supplier practice

There is a similar lack of data on the extent of industrial engineering in supplier countries. One investigation, which took a critical look at production management in apparel manufacturers in Commonwealth countries, found that almost everywhere industrial engineering capabilities needed to be upgraded (Lezama et al. 2004: 164) and that most suppliers in Bangladesh, Mauritius and South Africa reported the use of time studies to determine the SMV in order to evaluate direct labour costs rather than benchmark against external PTS standards (2004: 115).[9] In their study of Turkish knitwear suppliers, Bulgun and Vuruskan (2005: 10) found that many companies often had difficulty in archiving their manually undertaken cost estimates, and, in Cambodia, a USAID (2005:21) study on competitiveness of the apparel sector discovered that no factories were generating their own standard times.

Those suppliers that run their own industrial engineering departments and rigorously determine their labour costs in advance of CMT price negotiations may prefer an open book costing that adheres to a 'ball park' CMT price, lest they disclose sensitive costing data to the buyers. As a former international buyer and supplier has commented:

> Larger manufacturing groups have their own production engineers who will know the precise labour minute values on their garments, but may not disclose this to the buyer because they may seek to push this down.[10]

In discussing costing, the General Secretary of the Garment Manufacturers of Cambodia (GMAC) writes:

> The buyers nowadays come to us with the specifications of the garments they want produced ... what generally happens is that the factories are given a CMT price as a lump sum value and the factories

> are left to manage it as they like. In this case, there is some haggling and negotiations about providing a longer standard time in order to get a higher CMT price. In most cases however, there is just a negotiation to obtain a higher CMT price without much reference to the standard times. ... We are left to manage our own costs and the buyers generally adopt a take it or leave it attitude when it comes to the price they provide to us.[11]

Suppliers may, thus, end up selling to a buyer at a cost that is later found to be unachievable (before other unscrupulous buying practices are factored in), the consequence of which may be either a reduced wage bill or profit margin or both, and an unachievable delivery schedule based on the existing available capacity. This in turn can lead to low wages, excessive overtime or work undertaken in another factory – a move likely to constitute a breach of the commercial contract or code of conduct provision. In generally assessing wage problems along the supply chain, Vaughan-Whitehead (2010: 22) has identified that pay has commonly become 'the residual variable at the micro [factory] level' (see also Oxfam 2004; ETI 2007). Reliable empirical studies on purchasing practices are scarce, since suppliers are loathe to risk their commercial relationships by revealing the clauses inserted into contracts. However, anecdotal evidence and attempts by suppliers to fight back can shed some light on the situation.[12] In a telling and sober commentary, one expert stated:

> The wages paid to people on sewing machines in the developing world are an almost trivial part of a garment's CM cost. Overheads, ancillary work, finance, utilities, profit, 'legitimate' bribes (like paying customs officers who'd otherwise leave your exports to rot) and the owners' pension plans are typically several times the direct costs standard minutes measure. How a factory owner gets from those 6 minutes to a CM cost of $2.50 will vary massively from owner to owner, garment to garment, random moment to random moment (in Pakistan, the energy cost depends on whether the garment's going through when you can use gas from the grid or when you've got to use your own generator) and a gazillion other factors. A 10 per cent change in direct wages, or a 10 per cent change in the minutes needed, has an imperceptible effect: factory owners are hard-faced about paying rotten wages not because they're vital in themselves, but because they're just about the only cost factories can control – and in a world of fluctuating exchange rates, cotton price inflation,

arbitrary power disconnection and Western retailers going bust before paying suppliers, the only costs they can.[13]

Making human labour as important as fabric – towards a model

If buyers are serious about the wage commitments made in their codes of conduct or in the codes of multi-stakeholder initiatives to which they belong, then there is a responsibility to ensure that due provision is made in any CMT price to ensure that the prevailing wage – and where there is a code commitment a 'living wage' – is covered. If fabric can be costed out, and rendered as a dedicated and contractual cost item in the price negotiation of every garment bought, to what extent is it possible to determine and ring-fence the cost of labour in a price negotiation?

In order to generate a model for buyer use, it has proved necessary to critically examine predictive labour costing tools and conduct extensive interviews with staff from General Sewing Data (GSD), one of a number of companies specializing in the generation and application of standard times for costing in garment manufacture.[14] Two critical qualifications of predetermined time systems, in general, are necessary at the outset. The first relates to the validity of the claim that international standard times exist. The second relates to the scientific accuracy of systems based on synthetic times. GSD claims to be *the* international industry standard for the sewn products sector on the basis of a mention in the ILO publication, *Introduction to Work Study* (Kanawaty 1992: 426). There is also an assumption in academic circles that an apparel worker's effort should be fully contractible because 'international standards for the time it should take a skilled tailor to complete a particular task have been determined within the industry by time and motion studies' (Robertson et al. 2009: 10). Unfortunately, neither claim is correct. In the United Kingdom during the 1970s GSD began to apply methods time measurement to the whole range of sewing operations and became one of the foremost consultants in predictive labour costing with the majority of clients in manufacturing.[15] Without any comparison of standard allowed minutes across a sample of predictive labour costing consultancies for specific garment categories in controlled conditions, it is impossible to claim that there exist international standards in this area. Some buyers have called on the ILO to provide some leadership in this area.

Secondly, predictive labour costing prior to the placing of an order into production is by definition a synthetic undertaking and for this reason has always been looked upon with great cynicism, and at times disdain, by workers, and by trade unions in particular, as they have historically grappled with the contradictions of work measurement. The clothing workers of England and Scotland in the late nineteenth century suffered 'needless vexation' at the wide variations of the money (straight piece work) and time (payment on the basis of time saved) logs. The first national conference of the Scottish Amalgamated Society of Tailors and Tailoresses approved a uniform time log, which was submitted as a wage demand to the employers in 1866 (Stewart and Hunter 1964: 51). In the first quarter of the twentieth century, as scientific management grew apace, there was much debate and scepticism on the part of the trade union movement in particular concerning the extent to which detailed motion patterns could be standardized (Grant 1983), a fundamental issue in assessing the value of rate setting systems. In his seminal critique of works study, William Gomberg, the then Director of the Management Engineering Department of the International Ladies Garment Workers Union, challenged industrial engineers on the basis that all work standards were merely approximations and samples and were, therefore, not applicable to all workers on an equal basis with extreme mathematical and/or statistical accuracy. Gomberg (1948: 126) argued:

> while the mechanical and physiological factors may be controlled, the psychological factor, particularly under the impact of sociological pressures, leads to an unpredictable universe for all practical purposes.

To this would have to be added the impact of manufacturing conditions in often sub-tropical climes. These two qualifications notwithstanding, to what extent can some form of predictive labour costing assist what GSD refer to as *fact-based negotiation* during price/capacity negotiations between buyers and suppliers?[16] Since price negotiations may often take place away from the factory, such transactions should be guided by tariffs and standards as far as possible. As far as labour costing is concerned, this is a somewhat complex process. An internationally agreed tariff of standard allowed minutes for specific garments would only be achievable across different sites within a supply chain where the manufacturing process had been replicated identically in terms of methods, layout and machinery. For some large manufacturers this is not an impossibility, but the crucial variable for both the buyer and supplier is the difference between the SAM and what we can call the actual minute value in the manufacture of a garment under real factory conditions, otherwise called 'line efficiency'. In industrial engineering parlance the formula for efficiency is

$$\frac{\text{Output} \times \text{SAM}}{\text{Attended minutes}}$$

or the number of pieces assembled in the time available taking into consideration any 'downtime'. This is usually expressed as a percentage. The consensus in the industry is that an optimal manufacturing efficiency target should be 80–85 per cent. However, there are wide variations in many supplier countries. Two USAID Reports on Factory Competitiveness in Cambodia, for example, revealed variations in factory efficiency of between 35–80 per cent of international standard times (USAID 2005: 28; Nathan Associates, Werner International 2007). It is important to note that in work study parlance 'efficiency' is the responsibility of factory production management, with worker performance being but one of a number of variables (Bheda 2002: 8).

Unless the SAM for a specific garment takes on board the actual efficiency in a factory then code non-compliances are likely to be inevitable. Consequently a formula for calculating the actual minute value (AMV) of a garment needs to make allowance for variations in efficiency:

$$\text{AMV} = \text{SAM}/\text{efficiency} \times 100$$

Figure 4.2 shows a GSD costing sheet for the assembly of a basic adult five-pocket western-style jean indicating an SMV of 20.737 minutes. In our example here a figure of 27.642 minutes is allowed where efficiency is 75 per cent. So if the labour time input was costed at 75 per cent by a buyer but actual factory efficiency was running at only 50 per cent then factory management would have to cover the 25 per cent shortfall – either by subcontracting or excessive overtime with other attendant problems such as supervisory abuse or wage non-compliance. From a buyer's point of view they will be interested in a high level of efficiency and factory managers would be too if there was long-term stability, full order books and the prospect of taking more work on.

Style file number	**QJEANS001**
Style number / title	**5 Pocket Western Jean**
Style description	
Customer	
Season	
Product ref.	
Order quantity	
Comments	

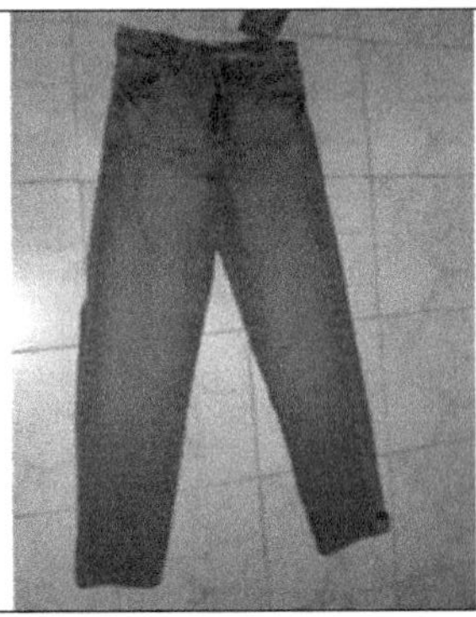

Created By	**GSD**	**08/02/2010**
Amended By	**GSD**	**22/02/2010**

	SMV	*@ Eff.*
Cutting	**2.496**	**3.327**
Machining	**15.110**	**20.141**
Examination	**2.161**	**2.881**
Pressing	**0.000**	**0.000**
Packing	**0.970**	**1.293**
Outwork	**0.000**	**0.000**
Total	**20.737**	**27.642**

Factory overheads	%	**0.000**
Fabric range variance	%	**0.000**
Learning curve allowance	%	**0.000**
Contract washing minutes		
Contract decorative minutes		
Total production cost		

Fusable	**0.000**	*Accessories*
Lining	**0.000**	*Packaging*
Fabric	**0.000**	*Embroidery*
Thread	**0.000**	*Misc*
Trimmings	**0.000**	
Total material cost		

Prime cost		
Social and employment costs		%
Sales and admin overhead		%
Contingencies		%
Quota		
Transport		
Freight insurance		
Total local cost		
Costing in	@	
Total cost		
Margin		%
Costed selling price		
Price point/actual margin		

Figure 4.2 Costing – standard minute value[17] for a five-pocket western-style jean

Source: Courtesy of GSD.

How might this facilitate a buyer in assisting a supplier to move towards a living wage?

Example: using predictive labour costing to move towards a living wage in Cambodia

Cambodia is a particularly instructive case. In 1999 the US–Cambodia Bilateral Textile Agreement linked quota access in the US market to factories' compliance with international labour standards, as monitored by the ILO Better Factories Cambodia project (now evolved into the ILO/IFC Better Work programme). The Cambodian apparel industry has proved resilient in the wake of the removal of the quota regime in 2004, but workers in the sector baulked at the declaration in 2010 of a new monthly minimum wage of US$61, engaging in a series of strikes to force the government and the employers to increase the national minimum wage to a living wage target of US$93 (Cambodia Institute of Development Study 2009a; Just-style 2010). Since the international competitiveness of the sector could potentially be compromised by a substantial hike in the minimum wage, the responsibility for such an increase must be shared by those buyers who choose to source from the country. An increase can only be achieved in any systematic ongoing way through the CMT, which in turn means that a calculation *will be required for each style ordered.* Let us consider how this might unfold in placing an order for the aforementioned western-style jean in Cambodia, were a buyer to make an effort to (a) ensure the payment of the prevailing wage and (b) assist the factory to move towards a living wage.

Ensuring the prevailing wage compliance cost

Working on the basis of a 26-day month, the number of available individual worker minutes would be 26 days × 8 hours × 60 = 12,480 minutes per month. A buyer must establish the existing prevailing wage and pay elements in each respective supply factory in Cambodia in order to determine the minute monetary value. The example here draws on factory wage data from the USAID Cambodian factory productivity study (Nathan Associates and Werner International op. cit.). In 2007 the average cost of labour was US$78.97 including hourly rate, benefits and overtime (see Table 4.1). However, in order to arrive at the basic earnings for calculation, it is necessary to subtract the overtime figure from the total to arrive at a standard remuneration figure for a basic week/month. In this case there is no incentive scheme, but other elements (attendance allowance, a seniority bonus and holiday pay) that make up the wage and would need to be factored into the calculation where they exist. Thus, our labour minute value would be:

US$78.97 – US$16.29 = US$62.68 (monthly take home excluding overtime) = 0.005 (US cents)

As we have seen already, however, factory efficiency levels in Cambodia can vary between 35 and 80 per cent. Thus, the actual standard allowed minutes could increase as follows:

Standard allowed minute	Actual minute value @ 80% factory efficiency	Actual minute value @ 35% factory efficiency
20.737	25.921	59.248

Thus, for the factory to be in a position to meet its prevailing wage compliance cost under a buyer code of conduct, the unit direct labour cost might vary between:

62.68/12480 = 0.005 (US$ cents)
× 25.921 (80% efficiency) = **13 US cents**

and

62.68/12480 = 0.005 (US$ cents)
× 59.248 (35% efficiency) = **29.9 US cents**

Determining a living wage costing

What would be the cost implications of moving to a living wage? If the monthly living wage figure of US$93[18] for Cambodia is divided by the available worker minutes in a month, one arrives at a figure of:

US$93/12480 = 0.75 cent per minute[19]

If this figure is then multiplied by the number of minutes required to make the garment at an agreed efficiency, then the unit labour cost would be:

AMV = 25.921 minutes × 0.75 (80 per cent efficiency) = 19.3 US cents

AMV = 59.248 minutes × 0.75 (35 per cent efficiency) = 44 US cents

Example: Consignment of 100,000 pairs of jeans from Cambodia

The existing direct labour cost for 100,000 pairs of jeans from a factory running at 35 per cent efficiency would be:

$$100{,}000 \times \text{AMV (29.9 US cents)} = \text{US\\$ } 29{,}900$$

The projected living wage direct labour cost in a factory running at 35 per cent efficiency would be:

$$100{,}000 \times \text{AMV (44 US cents)} = \text{US\\$}44{,}000$$

that is,

$$\text{US\\$}44{,}000 - \text{US\\$}29{,}900 = \text{US\\$ } 14{,}100 \text{ ('living wage supplement')}$$

Thus, the additional direct labour cost to a buyer for this consignment would be US$14,100.

Implications of the model

There are at least three areas where critical impact questions arise about the implementation of such a model for the sourcing company, factory management and the workers. These relate to the questions of cost, efficiency and determining a reliable mechanism for distribution. Making provision for living wage supplements would clearly involve an additional cost to the buyer who would have at least five (not mutually exclusive) options for dealing with this. Firstly, they could absorb the additional cost from their profit. This appears to be the approach currently adopted by Knights Apparel for their Alta Gracia factory in the Dominican Republic (Kline 2010). Secondly, they could pass the increase on to the consumer through a marginal increase in the retail price (Pollin et al. 2004).[20] This might be accompanied by a specific labelling initiative not unlike the current Fairtrade model (Nicholls and Opal 2005). Thirdly, they could absorb the increase but seek to pay for this through supply chain efficiencies rather than squeeze profit (ETI Purchasing Practices 2007). Fourthly, they could insist on the supplier absorbing the increase, in which case the factory management would have to absorb a reduction in profit to cover the increased labour cost. Fifthly, they could work with the supplier to increase productivity and efficiency to improve throughput and, by committing increased volume, enable the supplier to absorb this extra cost through factory

Table 4.1 Wage breakdown in a Cambodian factory

	Monthly base rate (US$)	Days/ month	Hours/ year	Daily	Hourly
a) Base rate	50.00	26	2,496	1.92	0.240
Attendance pay	5.00	–	–	0.19	0.024
Seniority pay	4.00	–	–	0.15	0.019
b) Base rate w/attendance and seniority	59.00	26	2,496	2.27	0.284
Holidays (23 days per year)	–	–1.9	–184	0.18	0.023
Vacation (18 days per year)	–	–1.5	–144	0.14	0.018
c) Adjusted for vacation and holiday	59.00	22.6	2,168	2.61	0.327
Overtime (1.5 × base rate × 2 hours)	16.29	5.6	542	0.05	0.007
d) Average monthly pay with overtime	75.29	28.2	2,710	2.67	0.333
Holidays 11.5 × 2 × base rate	3.69	1.0	92	0.04	0.005
e) Total including holiday compensation	78.97	29.2	2,802	2.71	0.338

Source: Nathan Associates/Werner International 2007.

efficiencies. Some UK fashion retailers, such as Marks & Spencer (Marks & Spencer 2010; ETI Wages Action Forum Notes 21 January 2010.), New Look (New Look Wages Project Update January 2010) and Asda George (2012) are already moving down this productivity path; and New Look has marginally increased its unit price by 4 US cents to assist one of their strategic suppliers to achieve this.[21] Where additional money is passed down to a supplier via the FOB price this could result in a degree of cost escalation as the consignment passes through the various stages of the value chain and additional cost elements calculated as a percentage exist (Miller and Williams 2008). Identifying and separately holding a 'living wage' premium on an account opened by the buyer could avoid extra cost in pass through, although separate payments outside commercial transactions may need special government clearance as in the case of Bangladesh.

Regarding efficiency, preliminary consultations with retailers about this model have generated concerns about varying factory efficiencies in any one country's supply base and that there would be no incentive on the part of a supplier to address efficiency if this was running lower than the level at which an agreed manufacturing cost had been calculated. One sourcing company as a matter of ethical buying principle does not insist on any disclosure of factory efficiency by way of a confidentiality agreement with its supplier. Others have become more hands on and have already been demanding greater efficiencies, urging their suppliers in longer term commercial relationships to sign up to in-country productivity initiatives, for example, in Bangladesh and Cambodia. Since manufacturing efficiency is a management responsibility and relates to optimum workstation lay-out, flow of materials and worker deployment (means, methods and materials) as well as worker performance, improved efficiency should impact on throughput in the first instance, and in turn on pace of work and line staffing levels. One value retailer, having identified its strategic suppliers of core products in Bangladesh and Sri Lanka has embarked on root and branch overhaul of their manufacturing lines under a lean manufacturing programme applying the same methodology across factories. This has led to worker up-skilling and upgrading in Bangladesh in particular, but also to a streamlining of assembly lines with improvements in efficiency and quality. Factory expansion has absorbed the redeployment of line workers (Asda George 2012).

Further research on the impact on workers of productivity-driven schemes is necessary, but what such initiatives reveal is a greater need for transparency between suppliers and buyers on manufacturing data. What is effectively 'open book costing' will require integrity measures on the part of buyers, such as long-term supply agreements, and the offer of productivity expertise where available. Ring fencing labour cost and calculating premia will require a significant adjustment by buying departments and probably a greater role for merchandisers. Software systems would need to be put in place to enable swift inputs of labour minute values, factory efficiencies and local remuneration packages. These detailed real-time monitoring systems are already in place in many fast-turn fast-fashion retail operations. One challenge is likely to be the impact of fast fashion retailing on the number of stock-keeping units (SKUs) and changing styles. However, in the same way that buyers and merchandisers have to keep track of material costs, so too should they maintain a database of minute values and labour costs for the factories in their supply base. For suppliers, what will clearly facilitate this process will be some familiarization if not standardization of work measurement and time analysis systems to achieve compatibility with those of their buyers.

One key question remains in relation to the reliable transfer of a living wage premium into the wage structure within a supplier factory. It is difficult to envisage how 'sustainable' labour costing can work without the issue of pay, effort and efficiency (systems of production management and workplace industrial relations) being addressed by collective bargaining, since the issue of equity looms large, and line efficiency initiatives are likely to be coupled with incentive schemes. However, in many supplier countries trade union density and collective bargaining penetration are extremely low or non-existent. Were a committed buyer to engage in the identification of a living wage premium, one way of addressing the industrial relations vacuum in the workplace could be the establishment of a dedicated account held by the buyer and the accrual of a premium based on subsequent orders during a financial year. This would yield an annual premium available for disclosure to the management and the workforce at an appointed time during the financial year. Where unions exist this premium could be the subject of an annual negotiation with the parties free to determine how the premium should be distributed via a collective agreement. Where unions do not exist and freedom of association has not been tested by the public declaration and dissemination to the workers of a non-victimization guarantee (ITGLWF 2010), workers could be invited to organize themselves to negotiate the distribution of such a premium on to the wage. The buyer could release the accrued amount on sight of a registered and auditable collective agreement from the said parties. However, since such a model could lead to the formation of 'yellow' unions, a variant might involve the differential from participating brands and retailers being deposited in a sectoral account to be the subject of a national collective agreement negotiated on the occasion of a national minimum wage determination.[22] A discussion of this variant, however, lies beyond the scope of this chapter.

Democratically elected trade union and employee committees are unlikely to be privy to the bargaining agenda likely to emerge in relation to any discussions about efficiency. Such understanding garnered during years of trade union workplace negotiation is in danger of being lost as those full time officers and trade union specialists in the developed (buying) countries retire. Workplace trade unions and employee committees will fail to satisfy the principles of freedom of association if they are prevented from affiliating with the wider trade union movement, where they will be able to access such expertise. There is clearly a role for organized labour here in relation to the servicing of the Better Work Performance Improvement Consultative Committees (PICCs) so that through training the PICCs agenda can move beyond production planning and control, quality assurance, technical and product design (Better Work 2011) to the issues of incentive pay and effort.

Given the fragmented nature of international apparel outsourcing and current interpretations in some quarters of anti-trust legislation, buyers undertaking such initiatives must – for the time being – do so in isolation. If only certain workers on lines producing for buyers committed to costing on the basis of a living wage were to receive the premium, this might have a de-stabilizing impact on industrial relations. In recent consultations with workers and management of two Bangladeshi factories where a single buyer declared its readiness to pay more, both parties insisted on the amount being paid to all workers at the facility in order not to compromise workplace equity.

Conclusions

Unless a sourcing company wholly owns a manufacturing facility, its ability to achieve wage compliance, raise wages and address excessive overtime in a competitive multi-buyer made-to-order production system will always be compromised by the extent of its stake in overall production and by the duration of the existing and planned commercial relationship with its supplier(s). Thus, for sourcing companies with living wage code commitments, the reality is that the nexus of wage compliance lies in the individual transaction of an apparel order. This does provide an opportunity for a buyer to *contribute to* the raising of wages through the negotiation of the manufacturing (CMT) cost. When allowance is made for factory efficiency, incentive and any existing local factory payment system, it is possible to calculate a labour minute value for any garment, 'ring fence' the agreed labour cost and make this an explicit part of the commercial contract between the buyer and the supplier, in the same way that fabric is itemized in FOB negotiations. The same methodology can also permit the identification of a 'living wage supplement' subject to consensus with national stakeholders, specifically the trade unions, regarding such a figure. Buyers can retain these supplements on account and pending disclosure of the amount to the workforce, and the negotiation of a collective agreement, release the funds to the factory. Where trade unions do not exist, workers shall be invited to organize themselves to access this premium. In this process, freedom of association and collective bargaining can be credibly and routinely audited. Such an initiative is likely to succeed in the first instance in those manufacturers that are considered to be strategic partners by their sourcing companies and manufacturing basic garments. Preconditions for such an initiative require an alignment of methodologies for calculating standard minutes and, where wider initiatives are broached across a particular supply chain, a common approach to the improvement of factory efficiency.

Better Work/ILO could play a role in four key ways here. Firstly, there is a need to determine the extent to which there is a degree of harmonization in SMVs generated by those companies offering predetermined time standards for the purpose of labour costing. It is currently a misnomer to refer to these as international standards. A blind review of values generated by companies for specific garments under the auspices of Better Work/ILO would assist this process. Secondly, Better Work could extend its existing cooperation with the global union covering the apparel sector on training materials for PICC members on incentive schemes. This is an area where there is extensive trade union experience (Abernathy et al. 1999; Carew 2003; Gomberg 1968). Thirdly, the creation of a positive environment for freedom of association and collective bargaining continues to remain a largely unaddressed item on the compliance agenda (ITUC, CCC, IndustriALL, UNI 2012). Better Work could facilitate the publication of right to unionize/non-victimization guarantees and union access agreements in the supplier factories of participating brands. Fourthly, since more medium- to long-term solutions to the problem of poverty wages in the apparel sector point to multi-buyer and state approaches in the future, the whole issue of the application of anti-trust in the field of compliance merits a thorough critical examination. This is an issue deserving of future ILO/IFC support.

Notes

1. The author wishes to thank ActionAid UK for providing financial assistance to undertake the preliminary research for the original (unpublished) discussion paper. I am grateful to GSD Corporate Ltd., in particular the managing director Paul Timson and operations manager John Dutton, for their assistance and data on SMVs on particular garments, which were invaluable in developing the model. Thanks are also due to Mike Flanagan of Clothesource who provided data on import prices and trends into the European Union (EU) on selected garment categories, and provided critical comments on the draft. The author would also like to thank the following individuals for their assistance: David Birnbaum, Jean Jenkins, Sean Chiles, Razaul Karim Bhuiyan, Auret van Heerden, Andre Kriel, Steve Grinter, Hans Wettengl, Ken Loo, Derek Boyden, Derek Cattell, Edgar Romney and those company representatives who provided information on their costing practices.

2. Every worker has a right to compensation for a regular work week that is sufficient to meet the worker's basic needs and provide some discretionary income. Employers shall pay at least the minimum wage or the appropriate prevailing wage, whichever is higher, comply with all legal requirements on wages, and provide any fringe benefits required by law or contract. *Where compensation does not meet workers' basic needs and provide some discretionary income, each employer shall work with the FLA to take appropriate actions that seek to progressively realize a level of compensation that does.* More information is available at: http://www.fairlabor.org/fla/go.asp?u=/pub/mp&Page=FLACodeConduct (accessed 31 August 2011).
3. Bringing the industry into minimum wage compliance has been one of the major if modest achievements of Better Factories Cambodia (cf. Synthesis Reports 1–26) but the persistence of short-term contracts has had a negative effect on overall wage compliance.
4. In June 2012 the ITGLWF merged with the International Metalworkers' Federation and the International Chemical, Energy and Mineworkers' Federation to form IndustriALL.
5. ITGLWF response to Transfair's proposal to pilot Fair Trade Certified apparel for the US market (4 February 2010).
6. For US companies the relevant legislation is the Sherman Anti Trust Act, 2 July 1890, ch. 647, 26 Stat. 209, 15 U.S.C. §§ 1–; For companies in the EU the relevant European statute is Article 101 of the Treaty on the Functioning of the European Union (TFEU), and the respective national laws governing competition policy for each member state. Cf. summary of advice given to the UK Ethical Trading Initiative by Rupert Anderson QC and Alan Bates of the London Bar in 2008 in respect of the compatibility with United Kingdom and EU law of a proposed scheme to encourage manufacturers in Bangladesh of garments and other goods destined for sale by retailers in the United Kingdom, to pay workers a 'living wage'. Cf. also the advice provided by DLA Phillips Fox to Oxfam Australia in respect of the Global Play Fair Alliance campaign to improve the wages and conditions of workers in the sportswear industry (Oxfam Australia 2011).
7. Also known as predetermined motion time systems (PMTS).
8. Email correspondence with a sourcing director at an ETI member firm.
9. One leading high street retailer in the UK reported that out of 1,300 of its apparel suppliers, it knew of only 20 that were using GSD.
10. Interview with former international buyer and production manager.
11. Email correspondence with Ken Loo, General Secretary of the Garment Manufacturers Association of Cambodia, 7 December 2009.
12. See Forum of Private Business Late Payment Survey 2008 http://www.fpb.org/images/PDFs/surveys/FPB%20-%20late%20payment%20survey%20results_Aug%202008.pdf (accessed 19 February 2010).
13. Email correspondence with Mike Flanagan of Clothesource.
14. GSD provided a set of SMVs on a range of staple fashion garments and assisted with the development of the costing methodology. In order to calculate an example, actual labour costs were necessary, which again, for reasons of commercial sensitivity, are not readily available. However, some useful data were publically available from the value chain work undertaken in Cambodia by Nathan Associates and Werner International (2007).

15. Other commercial consultancies that compete in this field are MODAPTS, Seweasy, ProMan and Sewing Performance Data.
16. Interview with Paul Timson of GSD.
17. GSD use the term standard minute value to denote the standard allowed minute on a specific garment.
18. This is the lower figure suggested in CIDS (2009a).
19. Rounded to the nearest decimal point.
20. There is some debate as to how marginal this increase would be. Percentage add-ons as the garment passes through intermediary stages before landed duty paid stage might inflate the FOB (Miller and Williams 2009).
21. Letter to ActionAid dated 19 January 2010. Significantly, Marks & Spencer, which has been working with GSD with its model factories in Bangladesh, has announced in its Plan A commitments for 2010 to 2015 to:

    > Implement a process to ensure our clothing suppliers are able to pay workers a fair living wage in the least developed countries we source from, starting with Bangladesh, India and Sri Lanka by 2015. We will achieve this by ensuring that the cost prices we pay to our suppliers are adequate to pay a fair living wage and by rolling out our ethical model factory programme to ensure the cost price benefits are paid to workers.

    http://corporate.marksandspencer.com/documents/publications/2010/planacommitments2010 (accessed 31 August 2011).
22. I am grateful to Jim Baker (ITUC) for this observation and contribution.

Part II

5
What Does 'Fast Fashion' Mean for Workers? Apparel Production in Morocco and Romania[1]

Leonhard Plank, Arianna Rossi and Cornelia Staritz

Introduction

The expansion of organizationally fragmented and geographically dispersed global production networks (GPNs) has been an important source of employment generation in many developing and transition countries over the past decades. However, the quality of employment generated in these GPNs is often characterized by a high degree of flexibility, uncertainty and precariousness. These employment quality characteristics may be specifically relevant in the increasingly important fast fashion segment of the apparel industry. At the heart of the fast fashion strategy lies a business model based on increased variety, flexibility and permanently shrinking product life cycles that require bringing new products to markets at an increasing pace and in shorter time spans. This implies not only increased organizational flexibility and shrinking lead times for supplier firms but also delivering high quality apparel items at low cost. In this context, supplier firms may struggle to accommodate potentially conflicting requirements and may pass on these pressures in the form of expanded and flexible work hours, intensified production processes and delayed wage payments to their workforce. These pressures may be particularly intense for regional suppliers located in countries in geographic proximity to the key end markets of the EU-15, the United States and Japan that derive their competitive advantage from being integrated into the fast fashion segment of apparel GPNs.

Against this background, this chapter assesses how integration into GPNs in the fast fashion apparel segment is impacting workers and social upgrading. It particularly analyses whether the sourcing practices related to fast fashion, such as short lead times, high flexibility, speed of production, low costs and high quality, create additional hurdles to social upgrading. The focus is on the apparel industry in Central and Eastern Europe and North Africa with comparative case studies from Morocco and Romania, due to their importance as regional and fast fashion suppliers for Western European lead firms.

The chapter relies on trade and national industry data as well as on primary data collected in 2008 through firm-level semi-structured interviews with managers and workers' representatives, standardized interviews with workers in Romania and focus group discussions with workers in Morocco. The non-random samples were selected based on differences in geographical location within each country, firm size (number of employees), positioning in GPNs and institutional specificities (former state-owned versus greenfield location in Romania; participation in the *Fibre Citoyenne* social compliance initiative in Morocco). More specifically, in Morocco, a sample of 19 firms located in the main four industrial poles – Casablanca, Rabat, Fes and Tangier – were selected. Among these, 13 firms are suppliers for fast fashion retailers such as Zara and Mango.[2] In Romania, a sample of 12 firms that largely supply fast fashion products to retailers such as H&M, C&A and Zara geared towards the EU-15 market was selected. Most of these firms are concentrated around Bucharest as well as in the south and the north-east of Romania.[3] Key informant interviews with stakeholders in the apparel industries in Morocco and Romania, including the government, employers' associations, trade unions, non-governmental organizations (NGOs) and industry experts complemented the firm-level interviews.

The section below outlines the concept of social upgrading in the context of the production trends brought about by the fast fashion model. The following section focuses on global and macro-regional dynamics in the apparel industry in the region with a focus on the emergence of fast fashion, regional trade agreements and the development of the apparel industries in Morocco and Romania. The 'social upgrading in fast fashion' section discusses the empirical findings with regard to the roles of workers and social upgrading experiences in fast fashion apparel GPNs highlighting similarities and differences between the two case studies. The last section concludes by summarizing the main findings and identifying policy areas to further economic and social upgrading.

Social upgrading in fast fashion

Evidence on the outcomes for workers as a result of insertion in GPNs is mixed. On the one side, GPNs have undoubtedly created new employment opportunities for previously marginalized groups of workers, such as women and unskilled workers, who did not have access to this type of wage employment beforehand. At the same time, globalization of production and especially commercial requirements deriving from buyers' purchasing practices and their search for low costs often lead to high pressures on supplier firms. These pressures are often associated with low labour costs and an increase in flexible and vulnerable labour arrangements, such as temporary, contract and migrant labour. For these categories of workers, participation in GPNs may have led to increased vulnerability and insecurity (Barrientos et al. 2011; Carr and Chen 2004; Chen et al. 1999; Collins 2003; Hale and Wills 2005; Knorringa and Pegler 2006; Oxfam International 2004; Raworth and Kidder 2009; Standing 1999).

Commercial pressures radiating from lead firms include not only low cost but also high quality, short lead times, responsiveness to changes in orders and flexibility. These are related to lean production and just-in-time delivery, which enable retailers and manufacturers to maintain low inventories of goods thereby reducing storage costs and allowing rapid and flexible response to changing market conditions. In contrast to standard conceptions of flexible labour markets where labour costs reduction drives flexibility, employment strategies tend to be more nuanced within GPNs because of the need to balance cost reduction and flexibility on the one side with consistency and quality of supply that meets lead firms' standards on the other side. Production for GPNs has, therefore, been linked to the emergence of a new type of labour, which is characterized by a high degree of flexibility and cost effectiveness but also meets quality standards (Barrientos et al. 2011).

These contradictory demands from lead firms are particularly pronounced in the increasingly dominant fast fashion segment in the apparel industry into which suppliers in Central and Eastern Europe and North Africa are often integrated due to their geographical proximity to Western European end markets. The fast fashion segment of the apparel industry, pioneered by the Spanish retailer Inditex (particularly with its Zara brand) is renowned for the ability to make cutting-edge fashion trends accessible to high street customers in a matter of weeks. Fast fashion is, therefore, characterized by short product life and production cycles, responsiveness to the latest fashion trends and speed and flexibility of production, as well as consistency and quality that are as crucial as costs to remain competitive.

It is therefore important to investigate how these complex commercial dynamics in the fast fashion segment of apparel GPNs impact workers and what the stimuli and constraints are for social upgrading. Social upgrading is composed of *measurable standards* such as wages, physical wellbeing (e.g. health and safety, working hours) and employment security (e.g. type of contract, social protection); and *enabling rights* such as freedom of association and collective bargaining, the right to freely choose employment, non-discrimination and voice. Enabling rights are critical to improve labour standards by empowering workers to engage in an effective relationship with their employers (Elliott and Freeman 2003).

Social upgrading is not linear as workers are not a homogenous group. Some workers may experience social upgrading whilst others may be left behind. The different impact can be related to the status of workers – regular vs. irregular workers or workers in core plants vs. subcontracting plants – as well as to their gender and ethnic background (Barrientos et al. 2011; Barrientos and Smith 2007). Where irregular workers are over-represented by women, ethnic and migrant groups, they often face double discrimination (both through their social and employment status).

The apparel industry in Central and Eastern Europe and North Africa

Fast fashion and regional suppliers in Central and Eastern Europe and North Africa

In the past three decades, the apparel industry in Central and Eastern Europe and North Africa has experienced dramatic transformations in the context of European macro-regional integration. The deepening of regional production networks has been propelled by changing industry dynamics and corporate strategies, in particular the increasing importance of time in sourcing decisions and the emergence of fast fashion, as well as the macro-regional integration process driven by RTAs (Pickles 2006).

One of the most influential trends in sourcing strategies of apparel lead firms is the increasing importance of time. This is related to the shift to lean retailing and just-in-time delivery where buyers defray the inventory risks associated with supplying apparel to fast-changing, volatile and uncertain markets by replenishing items on their shelves in very short cycles and minimizing inventories (Abernathy et al. 1999, 2006). The increasing dominance of fast fashion underlines these developments. Retailers such as Inditex/Zara and H&M have come to be known as the avant-garde in this respect and have gained increasing shares of the world apparel market with Inditex becoming the world's leading specialized apparel retailer (*The Economist* 2005, 2008).

However, shorter lead times, quick response and flexibility have become important not only for fast fashion retailers. While only a few retailers can be labelled genuine fast fashion retailers where their entire sourcing strategy is geared towards fast fashion, many traditional retailers that used to have four seasons per year, such as Marks & Spencer, C&A, Next, New Look, Gap and The Limited, follow fast fashion sourcing strategies for their more fashion sensitive product lines (Plank and Staritz 2011; Rossi 2011). Most recently, even luxury labels such as Chanel and Gucci are under pressure to offer new products more often (Roberts 2010). Hence, fast fashion sourcing trends have become increasingly important across the apparel industry. One consequence of this development is that geographic proximity to end-markets has become an important factor in sourcing decisions.

Another influential trend in buyers' sourcing decisions is the increasing importance of labour compliance in response to pressures from campaigns by NGOs and compliance-conscious consumers. As large retailers are often highly recognizable high-street brands, they are particularly sensitive about their brand reputation and have invested in corporate social responsibility (CSR) and developed codes of conducts (Jenkins et al. 2002; Locke et al. 2007; Mamic 2003; O'Rourke 2002). There is often an intrinsic tension between the commercial pressures with regard to low costs and high flexibility and the CSR requirements imposed by buyers, which by default are likely to increase the cost of production and reduce flexibility.

Organizational dynamics in apparel GPNs also have to be assessed in the context of the changing regulatory landscape. Macro-regional integration has played an important role in the furthering of regional production networks. Special trade agreements – referred to as outward processing trade (OPT) – created favourable conditions for the offshoring and outsourcing of labour-intensive production processes to nearby countries as part of a broader strategy to secure the competitiveness of the European apparel and textile complex (Dicken 2003; Gereffi 1999). This was achieved by allowing EU-based firms to temporarily export inputs for processing to an OPT-partner country and re-import products under preferential conditions, that is, only paying duty on the minimal value-added (labour) taking place in the neighbouring country (Pellegrin 2001). In the case of apparel, it generally involved the export of EC/EU inputs (fabric, cuttings or semi-finished apparel) to nearby lower-cost countries, which made them up into ready to wear apparel for re-import. These trade arrangements promoted a specific division of labour where low cost regional neighbours were largely responsible for labour-intensive assembly production whereas more capital-intensive and higher value activities remained based in the EC/EU. As integration deepened in the context of EU accession or the Euro-Mediterranean Partnership, these specific rules of origin regulations were expanded, but production structures tended to be sticky due to a deep seated division of labour based on OPT relationships (Begg et al. 2003). These OPT arrangements laid the ground for a flourishing intra-regional apparel trade in the 1980s and particularly in the 1990s. Western European apparel manufacturers and retailers increased their involvement in the region, with specific sourcing arrangement being based on geographical location, cultural affinity, historical factors, national industry pressures and existing structures and business contacts (Begg et al. 2003; Pincheson 1995; Textiles Intelligence 1997). In the context of regional trade agreements and fast fashion, regional supplier countries increased their market share in the EU-15 in the 1990s and early 2000s. The boom in apparel exports from the region lost momentum particularly at the end of 2004 with the MFA phase-out, as orders shifted to China and other low-cost Asian apparel exporter countries (Frederick and Staritz 2012; Gereffi and Frederick 2010; Staritz 2011). However, these reductions have not been as dramatic as expected. The still important role of regional supplier countries for the EU-15 is revealed in Table 5.1. Countries from Central and Eastern Europe and North Africa accounted for a market share of 25.3 per cent and 27.3 per cent in 1995 and 2004 respectively, which declined to 19.5 per cent in 2011.

Fast fashion and Morocco's and Romania's apparel industries

Morocco and Romania are key regional and fast fashion suppliers to the EU-15 market (ranking eighth and seventh in 2011 respectively, Table 5.1).

In Morocco, the apparel industry has been at the forefront of the export-oriented development model that has been embraced since the first half of the 1980s (Cammett 2007), when apparel exports increased significantly. While apparel exports accounted for 4.5 per cent of total merchandise exports in 1980, they increased to 32.2 per cent in the early 2000s. Formal employment grew throughout the 1990s and early 2000s peaking in 2003 with a registered workforce of 168,480. OPT relations – known as *admission temporaire* in Morocco – had an important role in promoting integration into production networks of European retailers and manufacturers (Cammett 2007; GATT 1989; Staritz and Frederick 2012). By the late 1990s, the share of apparel exports using temporarily imported inputs for re-export under this regime reached 90 per cent (GATT 1989; ILO 2000).

Table 5.1 Top 10 apparel exporters to the EU-15

	Value (in million EUR)						Market share (%)					
	1995	2000	2004	2008	2010	2011	1995	2000	2004	2008	2010	2011
World	**50.377**	**78.117**	**85.393**	**103.758**	**105.484**	**114.131**						
EU-15	21.838	30.513	32.642	38.812	37.887	40.862	43.3	39.1	38.2	37.4	35.9	35.8
China	3.542	7.450	11.038	24.330	27.539	28.661	7.0	9.5	12.9	23.4	26.1	25.1
Turkey	3.189	5.322	7.520	7.612	7.610	7.923	6.3	6.8	8.8	7.3	7.2	6.9
Bangladesh	967	2.567	3.689	4.667	5.748	7.363	1.9	3.3	4.3	4.5	5.4	6.5
India	1.588	2.005	2.434	3.826	4.135	4.533	3.2	2.6	2.9	3.7	3.9	4.0
Tunisia	1.729	2.567	2.586	2.580	2.312	2.401	3.4	3.3	3.0	2.5	2.2	2.1
Romania	972	2.558	3.679	2.349	1.958	2.310	1.9	3.3	4.3	2.3	1.9	2.0
Morocco	1.631	2.356	2.417	2.386	2.084	2.146	3.2	3.0	2.8	2.3	2.0	1.9
Poland	1.604	1.826	1.153	1.421	1.830	1.993	3.2	2.3	1.4	1.4	1.7	1.7
Viet Nam	–	–	–	1.201	1.337	1.646	–	–	–	1.2	1.3	1.4
Greater Europe*	**12.746**	**20.599**	**23.330**	**22.136**	**20.819**	**22.252**	**25.3**	**26.4**	**27.3**	**21.3**	**19.7**	**19.5**
MENA-4**	3.508	5.222	5.351	5.451	4.808	5.004	7.0	6.7	6.3	5.3	4.6	4.4
CEE***	6.049	10.055	10.459	9.073	8.401	9,324	12.0	12.9	12.2	8.7	8.0	8.2
Turkey	3.189	5.322	7.520	7,612	7,610	7,923	6.3	6.8	8.8	7.3	7.2	6.9

Note: * MENA-4, CEE and Turkey. ** MENA-4: Egypt, Jordan Morocco and Tunisia; *** CEE: Albania, Bosnia and Herzegovina, Belarus, Bulgaria, Croatia, Czech Republic, Estonia, Hungary, Latvia, Lithuania, Republic of Moldova, Montenegro, Poland, Romania, Serbia, Slovakia, Slovenia, the former Yugoslav Republic of Macedonia, Russian Federation and Ukraine.

Source: Eurostat; apparel represents HS61+62; world value represents the sum of EU-15 intra and extra trade; (–) indicates country not in the top 15 in given year.

In Romania the apparel industry had an important role in the industrialization process under state socialism (Begg et al. 2003). Alongside the overall economic downturn, production in the apparel industry declined sharply after 1989. However, the apparel industry recovered quickly due to OPT relationships with Western European firms. The OPT contracts – called *Lohnsystem* in Romania – provided orders as well as materials and machinery firms could not finance otherwise. However, they also established a division of labour that furthered the disintegration of the domestic textile and apparel complex at the industry level and a change from integrated full-package production to assembly manufacturing, involving a shift to cut, make and trim (CMT) production at the firm level, while sustaining and expanding employment at low wages. The apparel industry developed into a major pillar of the Romanian economy absorbing almost a fifth of total industrial employees and accounting for more than a quarter of total exports in the early 2000s; around 80 per cent of these exports occurred under the *Lohnsystem* (Plank and Staritz 2011). Formal employment in the apparel industry reached its peak in 2003 where it accounted for 303,000 workers.

In the context of the MFA phase-out at the end of 2004, exports declined in both countries as global buyers shifted orders towards low-cost Asian suppliers. However, while exports and employment continued to decline in Romania, the Moroccan apparel industry stabilized in 2006, in particular through orders from fast fashion retailers from Spain (Table 5.2). Retrenchment and consolidation in Romania were more pronounced as firms were not only confronted with increased competition related to the MFA phase-out but also had to face rising production costs in the context of EU accession. The remaining firms tried to move away from the increasingly precarious *Lohnsystem* by taking on more functional responsibilities, diversifying to the domestic and regional markets and reducing (labour) costs through internal relocation of production to poor regions within Romania and to neighbouring non-EU countries (e.g. the Republic of Moldova and Ukraine[4]) (Plank and Staritz 2009, 2011). Notwithstanding the critical situation brought about by increased competition post-MFA, the apparel industry remains important in both countries, in particular for employment generation. In 2007, the apparel industry accounted for 42 per cent and 15 per cent of total manufacturing employment in Morocco and Romania respectively (Table 5.2). In both countries, the workforce is characterized by a high share of female workers, with women accounting for 84 per cent and 87 per cent of the total workforce in 2008 in Morocco and Romania respectively (Plank and Staritz 2011; Rossi 2011).

Table 5.2 Key indicators of Romania's and Morocco's apparel and textile industries

	1991	1995	2000	2003	2004	2005	2006	2007	2008	2009	2010
Romania											
Apparel exports (US$ mils.)	262	1.396	2.737	4.828	5.374	5.177	5.079	4.394	4.231	3.218	3.327
Annual growth (%)	n.a.	36	7	21	11	–4	–2	–13	–4	–24	3
EU-15 share (%)	91	94	91	90	90	88	87	84	85	84	84
EU-15 unit value (€)	n.a.	18,0	20,0	20,4	20,9	22,4	24,0	43,2	36,8	25,7	n/a
Textile imports (US$ mils.)	183	977	1.563	2.706	3.184	3.153	3.183	3.108	3.157	2.415	2.681
EU-15-share (%)	76	86	88	85	83	79	77	72	73	72	71
Employment apparel (ths.)	244	189	261	303	281	259	247	215	193	173	n.a.
Annual growth (%)	n.a.	–9	9	0	–7	–8	–5	–13	–10	–10	n.a.
Share in manufacturing (%)	8	9	17	19	19	18	18	15	14	15	n.a.
Employment textile (ths.)	394	185	95	84	78	67	65	58	54	28	n.a.
Morocco											
Apparel exports (US$ mils.)	398	2.250	2.444	3.217	3.480	3.331	3.593	4.239	4.473	3.599	3.766
Annual growth (%)	n.a.	23	–3	14	8	–4	8	18	6	–20	5
EU-15 share (%)	91	97	94	94	93	93	90	89	88	87	85
EU-15 unit value (€)	n.a.	17,2	17,3	17,8	18,0	18,3	19,4	20,8	22,0	21,8	21,4
Textile imports	413	1.297	1.359	1.803	1.963	1.930	2.093	2.434	2.567	2.144	2.361
EU-15 share (%)	82	91	90	85	83	80	77	76	74	69	67
Employment apparel (ths.)	81	n.a.	157	168	162	150	152	153	n.a.	n.a.	130
Annual growth (%)	n.a.	n.a.	n.a.	n.a.	–4	–7	1	1	n.a.	n.a.	n.a.
Share in manufacturing (%)	n.a.	n.a.	43	47	n.a.	n.a.	n.a.	42	n.a.	n.a.	n.a.
Employment textile (ths.)	75	n.a.	49	n.a.	n.a.	n.a.	n.a.	39	n.a.	n.a.	34

Source: UN Comtrade (apparel exports and textile imports); Eurostat (EU-15 unit values); Employment Romania (NIS, 2008, 2009, 2010); Employment Morocco (HCP 2010; GoM 2011; MICNT 2011).

Both countries are predominately integrated in the fast fashion segment of Western European apparel GPNs. While the dependency on the EU-15 market has decreased over the first decade of the 2000s, it still remains around 85 per cent in 2010 (Table 5.2). A disaggregated view of the EU-15 end market reveals that Romanian and Moroccan apparel firms serve different national end markets. Morocco's main end markets are France (31.3 per cent) and Spain (30.1 per cent), while Romania's exports are concentrated in Italy (30.9 per cent) and Germany (20.1 per cent). These different end markets are related to historic and cultural affinities as well as to the time of insertion in the OPT regime. The different end markets are also reflected in the major buyers. The main buyers active in Morocco are Spanish (fast fashion pioneers such as Inditex/Zara and Mango), French (such as Decathlon) and British (such as Marks & Spencer) (Rossi 2011). In Romania, the fast fashion retailer H&M has had a prominent role employing around 10 per cent of the Romanian apparel workforce in 2006 (H&M 2006: 8). German buyers with a long sourcing history include C&A, KardstadtQuelle and Steilmann, as well as the British Marks & Spencer. Prominent branded manufacturers from Italy and Germany include Dolce & Gabbana and Hugo Boss (Plank and Staritz 2011).[5]

Morocco's and Romania's apparel industries are primarily locally owned and consist predominantly of small- and medium-sized firms.[6] Besides some functional upgrading processes, the majority of firms continue to largely operate as CMT suppliers. In Morocco, most firms operate a production model named 'co-traitance', which is an intermediate form between CMT and full-package. Under this arrangement the supplier can be responsible for fabric suggestion or some design additions but the buyer is still responsible for inputs sourcing and the associated financial risks. In 2010, the share of CMT-like contractual arrangements was estimated to range between 50 and 70 per cent (GoM 2011). In Romania, the *Lohnsystem* accounted for around 75–85 per cent of total apparel production but its importance has decreased as buyers are increasingly demanding that suppliers take over input sourcing. However, given the limited linkages to textile suppliers in countries outside the EU-15 and limited access to finance, particularly smaller supplier firms seem to be still involved in CMT production. Hence, whilst declining, CMT-like production arrangements where EU-15 buyers organize or are at least involved in input sourcing and other non-manufacturing functions, such as design and product development, still account for an important part of the industry (Begg et al. 2003; Plank and Staritz 2011; Rossi 2011). This is crucial for social upgrading, as in this production model labour costs account for the largest cost component, since inputs are sourced and financed by the buyer. This limits the room for improvements in wages and working conditions and possibilities to reduce costs and increase speed and flexibility through supply chain management or other management improvements. At the CMT stage there is, however, still scope for improvements, particularly with regard to reorganization of production and improved human resource and time management, which is often untapped in Morocco and Romania due to limited production management and supervisory skills.

Although functional upgrading in design and input sourcing as well as backward linkages into textiles have remained limited,[7] the industries in both countries have experienced functional upgrading in finishing processes and product upgrading. Investments in finishing activities, including laundry, embroidery, patchwork and printing, are particularly necessary in the fast fashion segment where retailers source more complex products and require quick and flexible responses from suppliers, and they are reflected in the relatively high unit values of apparel exports from Morocco and Romania (Table 5.2).[8] In Morocco, unit values have increased with the rise of fast fashion indicating the importance of more sophisticated and higher quality products. The situation is different for Romania, where unit values were always relatively high given its long history of manufacturing comparatively sophisticated products for German and Italian branded manufacturers. Fast fashion retailers also source relatively sophisticated products from Romania, such as suits for H&M. The higher quality of apparel exports and the prevalence of CMT production together with the path dependency of production structures based on OPT-relations (despite the formal phase out of OPT at the end of the 1990s[9]) are also shown in the continuing importance of (high cost) EU-15 textile imports that accounted for 67 per cent and 71 per cent of total textile imports in 2010 in Morocco and Romania, respectively (Table 5.2).

Social upgrading in fast fashion: findings from Morocco and Romania

The apparel industries of Morocco and Romania provide good comparative case studies to analyse the implications of fast fashion for social upgrading in the context of regional suppliers. They also highlight the similarities and differences in outcomes that have emerged as a result of the specific institutional contexts of the two countries. The empirical evidence in both countries demonstrates that the highly volatile nature of the fast fashion segment and potentially contradictory pressures with regard to buyers' business demands for low costs, flexibility, high quality and consistency as well as labour standard requirements create constraints on supplier firms and workers in ways that circumscribe social upgrading.

There are important similarities between the two countries with regard to social upgrading that relate to fast fashion sourcing strategies. In particular, flexible and short-notice orders are a challenge for firms and workers that impact on working time and work intensity (i.e. increased pace and stress at work). Interviews with Moroccan managers show that effective and compliant human resource management is rendered very difficult because of the flexibility and short notice of orders. This translates into precarious work contracting, as on certain weeks or days when there may be slow production and no urgent orders, managers send workers home early in the day in order not to pay for them being idle. When orders arrive, they usually entail a sustained production pace involving long working hours. As a result, Moroccan managers state that the legal requirements in terms of maximum working hours and overtime are unrealistic and often not respected, and that they regularly resort to compulsory overtime. In Romania, overtime and work intensity issues are also related to fluctuating and short notice orders. During low order periods, workers are sometimes forced to work (and earn) less, whereas in peak times minimum targets are very high and can only be met either by work intensification through reengineering the production process — difficult to achieve in the short term — or by increasing work intensity or working time. Romanian managers and trade unionists acknowledge that the scope for calculating a 'realistic' target that allows for decent working conditions is often limited by the buyers' requirements in terms of delivery time and price. A recurring practice in Romania with regard to overtime is that suppliers maintain two sets of working time records – one 'official' that is compliant with national law and buyers' CSR demands and one 'internal' that records the effectively worked hours. Orders in both countries also largely involve small sizes and, thus, small production runs, which make it necessary to change the production set up regularly, sometimes within a week, leading to additional pressures on firms and workers with regard to working time and work intensity.

As the fast fashion strategy not only relies on flexibility, fast product delivery, short runs and low prices, but also on high quality, reliability, to a certain extent more sophisticated products and compliance with labour standards (Acona 2004), buyers' commercial requirements are particularly challenging to fulfil. In the particular case of Inditex in Morocco, factory visits and interviews with Inditex have shown that the most expensive, top-end Zara items are produced in Morocco, requiring embroidery, applications, patchwork and pleating (this is supported by Tokatli 2008). Similarly, given the long existence of the apparel industry in Romania and the inherited skilled workforce, Romanian apparel exports are characterized by comparatively sophisticated products for German and Italian branded manufacturers and retailers. This means that alongside ensuring fast delivery times and being responsive to buyer demands, suppliers must provide high quality production, further increasing the pressure on both management and workers. A manager in Casablanca effectively described the tension he faces by saying that there is the 'law of the Law' and the 'law of the client', and for the immediate survival of the factory, as well as the jobs of the workers, he would rather keep the prices low and fulfil the client's business requirements and avoid investing in social upgrading.

A way for suppliers in Morocco and Romania to cope with conflicting buyer demands for flexibility without compromising quality is the use of different types of workers in the factory. In Morocco, regular, higher-skilled workers, who guarantee high quality, coexist with irregular workers with temporary or casual contracts, who can be hired and dismissed with great flexibility as workload needs change (Barrientos et al. 2011). Regular workers are often senior workers who have worked in the factory for many years and have a good relationship with management. Irregular workers are usually young, unskilled workers just entering the labour market who find themselves in the last segments of the production chain such as finishing, packaging and loading trucks for exports. Working on these activities adds pressure on irregular workers because all the delays accumulated throughout the manufacturing process are absorbed at this level. In Romania, the hiring of regular and irregular workers in the same factory is not common; however, a widespread strategy by suppliers to cope with the same pressures is to subcontract orders to smaller firms and workshops in rural areas or smaller towns in the surrounding area in order to increase production capacity quickly and to buffer the insecurities and flexibilities of orders. The motivation was the same as in Morocco with regard to regular and irregular workers – to maintain skills and workforce stability and secure quality by using core workers in the main plants, but coping with costs and flexibility pressures and the volatility of orders by using a network of smaller firms for mostly lower value tasks where cheaper and more flexible labour could be tapped.

Different workers experience varying degrees of social upgrading. In Morocco, it was found that regular workers tended to benefit in terms of skills training in order to respond to specific product requirements. In terms of measurable standards, they may be socially upgraded as a result of buyers' or multi-stakeholder social compliance initiatives, especially in terms of correct payment of wages and social security coverage. A similar pattern is observable in Romania where improvements in working conditions have taken place in formal apparel firms in areas such as measurable standards, including occupational safety and health (OSH) measures and formal employment contracts. Nevertheless, pressure on working time and work intensity as a result of fast fashion orders has also affected regular workers or workers in formal firms. However, in both countries it is the irregular or subcontracting workers who are most adversely affected by short-term and fluctuating orders and who have experienced the lowest levels of social upgrading or even social downgrading. The demand for shorter lead times and just-in-time delivery practices mean that suppliers effectively discriminate against a group of workers by keeping them on temporary and casual work arrangements (within the core firm, or between the core and subcontracting firms), imposing highly flexible and at times long working hours and mandatory overtime.

Industry pressures in GPNs are mediated by local institutional structures and regulatory contexts that play an important role in shaping social upgrading prospects (Coe and Hess 2013; Pickles and Smith 2010). Moroccan and Romanian firms reacted in different ways to accommodate industry pressures. Due to Romania's state socialist legacy, national labour codes, labour inspectorates and trade unions had been in place as part of the institutional environment and they have had an active role in securing certain labour rights, particularly with regard to OSH standards and formal employment contracts. This legacy distinguishes Romania from other apparel exporters such as Morocco, where similar structures are often weak or non-existent. However, the legacy of state socialism not only entailed 'assets' but also 'liabilities'. In particular, trade unions' role as transmission belts for the communist party largely delegitimized them as effective agents of worker interests. After 1989, they were slow to adapt to changing conditions related to global production and fast fashion and to organize a feminized and fragmented industry. Moreover, trade unions were concentrated in larger, former state-owned firms and largely lacked representation in newly founded firms after 1989. EU accession also impacted on labour legislation and social upgrading prospects of Romanian apparel workers with varying effects, as Romania had a national labour code in place with high levels of workers' protection (Pickles and Smith, 2010). In areas such as discrimination legislation was improved, while in others, in particular with regard to flexibility of employment relationships and working time, labour legislation was weakened (Plank and Staritz 2011).

In Morocco, the labour code was revised in 2004 and is considered among the most progressive in the Arab world. However, its enforcement is challenging due to the lack of resources for effective labour inspections. Freedom of association is a problematic issue in the apparel industry, where unionization rates are around 3 per cent with largely male participation, not reflecting the composition of the labour force. At best, there is widespread mistrust on the managers' side towards unionization efforts. In 2003, the Moroccan Textile and Garment Industry Association unilaterally issued an industry code of conduct and a related label, *Fibre Citoyenne*, to be awarded to factories passing a social audit. This initiative became particularly relevant in 2007 when Inditex pledged to source exclusively from factories labelled *Fibre Citoyenne* as part of the International Framework Agreement signed with the International Textile, Garment and Leather Worker Federation (Pfeiffer 2007). While the *Fibre Citoyenne* has had considerable positive outcomes for workers in terms of measurable standards (especially on OSH regulations, social security registration and correct records of employment contracts and recruitment procedures), it still fails to have a significant impact on enabling rights (Rossi 2013), in part due to its lack of a collective bargaining clause. Also, regular workers are the beneficiaries of the changes brought about by the initiative, whereas the same benefits are largely off-limits for irregular workers, who continue to operate under casual, temporary contracts and often are paid below the minimum wage.

In Romania, there exists no such industry-wide code, but in both countries private codes of conduct have proliferated as most buyers have established and implemented their own codes. These codes tend to have led to improvements in working conditions focused on measurable standards such as better lighting, ventilation or ergonomic chairs that relate to process upgrading as they also increase productivity by a more efficient use of the 'human resource'. In contrast, issues that are in conflict with the prevailing business logic (e.g. living wages, working time, work intensity, union rights) remain contested. This trade-off is revealed by contradictory demands from and limited coordination between the buying and CSR departments of buyers: suppliers state that they are faced with tight price and delivery time demands from the buying departments, and with labour compliance demands from the CSR departments and auditors, who, however, do not have the means to reward suppliers for improvements (e.g. via higher prices or more stable contractual relationships) (see also Oxfam International 2004).

Although the co-existence of some type of regular and irregular workers plays a crucial role in Morocco and Romania to cope with conflicting industry pressures, the concrete arrangements on the ground differ. While in Morocco flexible workers are often recruited through informal arrangements with regular and irregular workers working together in one firm, informal employment within larger core firms has been limited in Romania given the stricter labour code and the role of labour inspectorates in enforcing formal employment contracts. To accommodate flexibility demands, larger suppliers rather draw on subcontractors whose workforce is cheaper and more flexible and where informal work arrangements are more widespread than in the core plants. This phenomenon has more recently also increased in importance in Morocco, especially in the Tangier area. In particular, the *Fibre Citoyenne* has led to increased subcontracting, because larger firms applying for the label are not able to achieve the production targets in the time frame required by their buyers without imposing excessive overtime and resorting to casual contracts. To avoid being in non-compliance with the *Fibre Citoyenne* code in their plants, suppliers have increasingly been outsourcing their production to smaller workshops or informal garage firms, which is, however, clearly in violation of the code.

From the analysis it appears that social upgrading is not linear and participation in the fast fashion segment of apparel GPNs geared towards the EU-15 market has delivered a mix of social upgrading and downgrading in Morocco's and Romania's apparel industries. Table 5.3 provides a summary of the empirical findings. It shows that social upgrading has been selective with regard to both the type of workers and the dimension of social upgrading. Improvements have largely remained limited to measurable standards, where a case for enlightened human resource practices can be made and, thus, where social upgrading has fulfilled a role in the competitive strategy of firms. However, issues of high importance for workers such as wages, work intensity, overtime and enabling rights such as union representation have remained contested, since they potentially conflict with the prevailing industry dynamics, particularly demands associated with fast fashion. There are also crucial differences between social upgrading for regular versus irregular workers as well as for workers in core and

Table 5.3 Social upgrading/downgrading in fast fashion

Type of worker		Social upgrading/downgrading dimensions			
		Measurable standards		Enabling rights	
		Accessible	Contested	Accessible	Contested
Type of worker	Regular/core firm workers	OSH standards, employment contracts, social security, training	Wages, overtime, work intensity	Union representation in former state-owned firms in Romania	Union representation in Morocco and in newly found firms in Romania
	Irregular/ subcontracting firm workers		Employment contracts, wages, overtime, work intensity		Discrimination, voice, union representation

subcontractor firms, who are often discriminated against with regard to types of contracts, payment of wages, employment security, work intensity and flexibility, and skill training opportunities. Hence, social upgrading prospects are strongly influenced by buyers' purchasing practices particularly with regard to fast fashion strategies. Specific types of work and working conditions play crucial roles in enabling and sustaining industry dynamics in GPNs, in particular with regard to flexibility and the conflicting demands for low costs and high quality, and where labour rights are in contradiction with these industry dynamics they remain contested.

Conclusions and policy issues

This chapter has analysed the outcomes of insertion in Western European apparel GPNs from the perspective of two regional supplier countries, Morocco and Romania, emphasizing the challenges with regard to social upgrading in the context of fast fashion. The main findings of this chapter can be summarized as follows:

Buyers' purchasing practices have a clear impact on social upgrading trajectories, with fast fashion trends posing intrinsic challenges: The fast fashion business model, relying on quick response to fashion changes and customer demands, flexibility of production, low costs and high quality poses particular pressures on workers in supplier firms. While EU-15-based retailers have been key in ensuring continued apparel exports and employment in Morocco and Romania and have furthered certain economic upgrading processes particularly in the areas of finishing activities and more sophisticated products, this business strategy has had mixed outcomes for the quality of employment. In particular, the short lead times and flexibility in orders that characterize fast fashion production have negative implications in terms of flexible work arrangements, unrealistic production targets, excessive overtime and high work intensity.

Social upgrading has been selective in the apparel industries in Morocco and Romania with regard to both the dimension of social upgrading and the type of workers: The evidence presented in this chapter shows that while some workers with regular contractual arrangements in core firms have experienced a certain degree of social upgrading in the form of skills upgrading and improvements in measurable standards, irregular workers or workers in subcontracting firms have been largely excluded from social upgrading opportunities or even experienced social downgrading, given their role as a buffer for cost and flexibility pressures. Furthermore, the attainment of social upgrading is limited also for regular core workers, since certain issues such as wages, working time, work intensity and union representation that challenge business strategies in the fast fashion segment continue to be contested.

Buyers' CSR demands exacerbate the existence of a parallel workforce: Buyers impose strict cost, speed, flexibility and quality requirements while simultaneously expecting compliance with labour standards from their suppliers. These CSR requirements are, however, often implemented without coordination with sourcing practices and without being effectively and systematically embedded in companies' sourcing strategies. Therefore, they seem to exacerbate the use of a tiered workforce, with regular workers in core firms being the target of CSR and improvements in labour standards, while irregular workers in subcontracting firms are not.

Local institutional structures and regulatory contexts mediate fast fashion pressures on social upgrading: The examples of Morocco and Romania show that local institutional structures and regulatory contexts mediate the pressures on workers deriving from fast fashion. While in Morocco irregular workers often exist alongside regular workers in the same factory, in Romania the parallel workforce has taken the form of subcontracting to small and micro firms, as employment contracts are more widely enforced by labour inspections in core firms. Also, the legacy of state socialism provided structures with regard to labour regulation, labour inspection and union representation that contributed to mediate industry pressures. In Morocco, the *Fibre Citoyenne* label has contributed to raising awareness about compliance and has increased monitoring activities. However, although these specific local institutional and regulatory settings have to a certain degree contributed to protecting regular and core firm workers, the most vulnerable groups, such as irregular workers in Morocco and workers in subcontracting firms in Romania, continue to be largely excluded by social upgrading efforts.

These empirical findings point to policy areas specific to the different actors in apparel GPNs, which can improve the position of supplier firms and workers and further social upgrading.

One important step for buyers is to align their core-business activities and sourcing practices with their ethical commitments, reflected in their CSR initiatives. It is crucial to reduce volatility and flexibility in supplier relationships and orders and align contractual expectations with decent wages and working conditions. Another limitation of buyers' codes of conduct is their often narrow approach to social upgrading focusing on measurable standards and quick fixes to visible problems without taking into account broader dimensions of social upgrading that perceive workers as social actors with enabling rights. Further, buyers need to abandon their unilateral initiatives and engage with local stakeholders, in particular trade unions, NGOs and labour inspectorates or other local regulatory agencies to reduce multiple standards and strengthen local capacities to enforce labour rights.

In the context of asymmetric relationships between buyers and suppliers, supplier firms should try to diversify and engage with more buyers and markets to reduce dependencies (Plank and Staritz 2011). Other end markets, including regional and local markets, might exhibit better growth and upgrading potentials and allow for more beneficial outcomes (Pickles and Smith 2010). With regard to social upgrading, participation in industry-wide codes or multi-stakeholder initiatives can help to increase the bargaining power of suppliers vis-à-vis buyers with the objective of mediating the tensions between sourcing and compliance requirements (Rossi 2011).

Government policies play a crucial role for economic and social upgrading, in particular industrial, trade, innovation, skill and labour market policies, which need to be aligned within a comprehensive industrial development strategy. In the context of cost competitiveness, governments may be unable or unwilling to enforce labour rights and further social upgrading. However, cost competitiveness is not the whole story, particularly for regional suppliers that are largely out-competed on cost by lower-cost suppliers in other regions. The strategic location of regional suppliers needs to be complemented with industrial policies to further competitiveness and economic upgrading and align these strategies with social upgrading.

Governments in the EU-15 can also directly influence upgrading outcomes; for instance, the significant buying power of public authorities could be used as a lever to influence buyers' behaviour and thereby impact on economic and social upgrading trajectories. Another important area for intervention relates to trade policy where positive market access incentives can be linked to labour compliance requirements or improvements in the area of labour rights in supplier countries. EU member states could also demand from buyers based in their authority to ensure at least certain minimum labour rights in their supply chains. This would reduce competition in labour standards between supplier countries.

Civil society actors, including unions and NGOs, should continue to pressure for decent working conditions in GPNs. There are a variety of approaches that are pursued to this end, including campaigns directed at buyers, fair trade and decent work initiatives, and cross-border labour dialogues. The collaboration of international and local actors in developing or transition countries should be further encouraged as an effective strategy. On the level of social dialogue, unions have a crucial role in supporting social upgrading and have to further adapt to global production arrangements, collaborate regionally and internationally with other civil society actors and represent and approach all types of workers, including regular and irregular workers. At the regional and international level, European and International Framework Agreements between transnational corporations and global union federations have been signed to extend cooperative industrial relations to the companys' locations outside the home country. Although their effectiveness depends on a number of factors such as the commitment of buyers and the capacities and political environment of the unions involved (Lukas et al. 2010), they are promising instruments to be embedded in a wider regulatory framework that strengthens their enforcement structures.

The empirical findings presented in this chapter shed light on some of the main challenges of fast fashion production for social upgrading in GPNs. Further research is needed to assess the impact of fast fashion business models on workers vis-à-vis other more traditional production models. This is particularly relevant in light of the increasing shift towards flexible mass production in all manufacturing sectors. Also, a deeper investigation of the governance dimension of social upgrading and an analysis of public, private and multi-stakeholder initiatives along GPNs, especially in the fast fashion context, would be required to better understand the potential for policy interventions at local and global levels.

Notes

1. The authors thank Stephanie Barrientos, John Pickles and participants at the 2011 Better Work Conference 'Workers, Businesses and Government: Understanding Labour Compliance in Global Supply Chains' for comments on an earlier version of the chapter. The views expressed in the chapter are those of the authors and do not necessarily reflect those of their institutional affiliations.
2. We differentiate between fast fashion retailers and buyers that cannot altogether be described as fast fashion retailers but follow fast fashion sourcing strategies for specific product lines. The first group of buyers includes Inditex/Zara, Mango and H&M; the second group includes a much larger group of retailers, including for example Marks & Spencer, C&A, Next, New Look, Gap and The Limited.
3. Apparel firms are also concentrated in the North-Western region of Romania; however, these firms are largely integrated into GPNs of branded manufacturers, predominantly from Italy and Germany, and not retailers. The interviews and this chapter focus on suppliers that are integrated in sourcing networks of retailers, particularly in the fast fashion segment.
4. Similar regional relocation patterns have been observed between Slovakia and Ukraine (Smith et al. 2008).
5. Different types of buyers are associated with different regions within both countries. In Morocco, Casablanca and Rabat are associated with British and French buyers while Tangier is characterized mainly by Spanish buyers (Rossi 2011). In Romania, the North-Western region is associated with Italian branded manufacturers while retailers source largely from Bucharest, the South and the East (Plank and Staritz 2011).
6. In Morocco, there were around 880 firms registered in 2007 (Rossi 2011). In Romania, the number of firms accounted for around 6,300 in 2008. However, the overwhelming majority of companies are small (1,270) and micro enterprise (4,199), and an important share of those may not be operational (Plank and Staritz 2011).
7. Romania had an important textile industry before 1989; also in Morocco there existed capacities in textiles before the turn to export-orientation. However, given the dominance of OPT contracts that relied on imported inputs, the existing textile capacities were not further developed. The constant decline in terms of employment in the textile industry in both countries underlines this stagnation (Table 5.2).
8. Interpretations of unit values have to be taken cautiously as they may reflect higher quality and more sophisticated export products as well as a loss in competitiveness related to increasing costs. In this regard it is useful to assess unit values together with market shares. In particular in Romania the increase in unit values after 2004 that coincided with shrinking market shares is also related to rising costs.
9. OPT regulations phased out in Romania at the end of 1997 in the context of EU accession; in Morocco in 2000 in the context of the Euromed Partnership.

6

Voting with Their Feet? Explaining High Turnover and Low Productivity in the Lao Garment Sector[1]

Richard Record, Stephanie Kuttner and Kabmanivanh Phouxay[2]

Introduction: research question and rationale

This chapter investigates working conditions in the garments manufacturing sector in Lao People's Democratic Republic. It is based on an analysis of primary data collected through a garment sector survey coordinated by the World Bank in 2011. The survey consisted of a representative survey of firm managers, focus group discussions among current and former garment workers, and key informant interviews in the garment sector. There had previously been only limited structured documentation of these issues in Lao People's Democratic Republic[3], and the survey sought to begin filling the gap.

In the chapter, we first seek to better understand the context and dynamics of high workforce departure rates and lower productivity in the Lao garment sector (as compared both with other sectors in Lao People's Democratic Republic and with neighbouring garment industries). Second, we attempt to assess how labour policies and practices in particular are related to these high turnover rates and lower productivity. Finally, we look into how these issues are perceived by both managers and workers in the industry as part of efforts to identify possible win–win opportunities for improvement. That is, identifying changes in policies and practices at the level of the firm, the industry and the public sector that could help the Lao garment sector increase both profitability and worker wellbeing.

The following section provides more background on the garment sector in Lao People's Democratic Republic, situating it within the global garments value chain, and in the Lao national context in terms of both business and labour market dynamics, and in terms of industrial relations and labour law development. The survey data and methodology section follows. The authors' analysis of survey results, contrasting the perspectives of managers and workers, constitutes the main contribution of the chapter. The penultimate section presents the results of regression analysis to further substantiate these claims. Finally, the last section concludes by summarizing the findings and offering reflections on areas of possible further research and analysis.

Background: Lao garment sector and workforce

Most garment firms in Lao People's Democratic Republic were established in the early to mid-1990s as manufacturers sought to establish a production base in a country that was not quota constrained and had relatively low labour costs. Today, while modest by international standards, Lao People's Democratic Republic's garment industry is the country's largest source of manufacturing sector employment – employing more than 20,000 workers (or about 1 per cent of the country's total labour force) – and is a significant contributor to the country's annual exports. Similar to workforce demographics in garment industries globally, most factory workers are young women between the ages of 16 and 25, about half of whom have migrated from regions outside of the capital, Vientiane (where almost all factories are located). Garment sector work remains one of the few opportunities available to young women with limited skills or education to access wage employment.

A severe contraction in the Lao garment industry was widely anticipated following the expiration of the Agreement on Textiles and Clothing in 2005, but the sector has managed to confound expectations by surviving and, indeed, continuing to grow. As of mid-2011, garment producers in Lao People's Democratic Republic were reporting strong orders with a number of firms now unable to meet demand from buyers. Though the industry has managed to survive the international liberalization of the garment trade, it struggles to compete with neighbours given low productivity, high transport costs and long lead times associated with a landlocked supply chain.[4] Only the foreign-invested segment of the industry is well integrated into global value chains and able to exercise significant control over the supply chain. Indeed, findings from a recent global study of the post-MFA apparel industry would suggest that the small Lao industry will continue to face downward wage pressures from larger competitor countries unless it can successfully upgrade its apparel sector, including through investment in workers (Lopez-Acevedo and Robertson 2012).

Most garments produced in Lao People's Democratic Republic (around 75 per cent) are exported to the European Union (EU), due primarily to the additional tariff advantages associated with Lao People's Democratic Republic's duty-free access to the EU market under the terms of the 'Everything But Arms' Agreement, but also due to the small size of Lao garment producers and their inability to meet the larger orders required to compete in the US market.[5]

The Lao Government has also struggled to adapt from its role as 'central planner' to more one of 'regulator and facilitator'. This is evident in the area of industrial relations, which have not evolved apace with the emerging private sector and lag even in comparison to other regional 'transition' economies. As in both China and Viet Nam, Lao national trade union structures are fully incorporated into the political structure of the communist party (Yoon 2009). The representation of workers in the Lao garment sector and all other sectors of the economy remains mandated solely through the official mass organizations of the Lao People's Revolutionary Party (namely, the Lao Federation of Trade Unions, the Lao Women's Union and the Lao Youth Union).

While the 2006 Lao Labour Law establishes basic national labour standards (such as minimum wages, restrictions on overtime and entitlements such as maternity and sick leave)[6] and incorporates elements of International Labour Organization (ILO) core labour standards,[7] these appear to be only weakly enforced and there is very limited opportunity for third party arbitration or dispute resolution.

Labour standards established under contracts with foreign buyers or international trading regimes seem to create stronger inducements for compliance, although these systems of voluntary certification and periodic auditing do not appear to be sufficient to counter some of the unfair labour practices and difficult working conditions reported by workers. Absent is a system of collective bargaining or effective representation through official mass organizations, garment workers have limited effective bargaining power. As a result, when individuals are dissatisfied with working conditions, they have little alternative than to simply quit.

Survey data and methodology

The analysis presented in this chapter is based on primary data collected as part of a World Bank study between March and May 2011.

Structured firm surveys were conducted with owners and managers of garment factories using stratified random sampling of more than half of all firms, representing over 70 per cent of the garments workforce in Lao People's Democratic Republic. The surveys collected data on basic firm characteristics and workforce demographics, as well as on a range of labour standards and practices, including human resource management practices (recruitment, training, retention rates, etc.), labour relations and standards (health and safety, worker representation, etc.), labour productivity (rating labour-related constraints, use of performance indicators, etc.) and business practices and prospective. For the purpose of the survey, firms were grouped according to size: small firms (with less than 100 employees), medium firms (with 100–499 employees) and large firms (with 500 or more employees).[8] Figure 6.1 describes these size categories in terms of their share of all Lao garments firms (share of total firms) as well as their share of the total garments workforce (workforce shares). Actual numbers of firms observed in each size category are indicated in brackets. Thus, as can be seen, large firms constitute only 10 per cent of all firms, but almost half of the total garments sector workforce; medium firms employ another 43 per cent of the workforce; while small firms employ only 8 per cent of the workforce, although they number more than half of all firms.

Focus group discussions (FGDs) were conducted with garment factory workers from small, medium and large factories of both Lao and foreign ownership/management. Participants were selected to represent a range of demographic characteristics and work experiences (recent recruits and more experienced workers, rural migrants and Vientiane residents, different marital/family status, etc.). Additional FGDs were

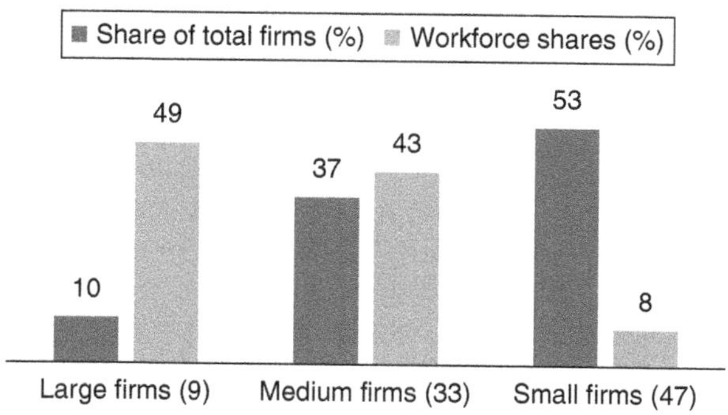

Figure 6.1 Lao garment firm and workforce shares
Source: World Bank Lao garment sector survey (2011).

conducted with male garment workers as well as with former workers. Discussions were structured around the following topics: recruitment experiences and expectations; working and living conditions; workers' social status and expectations; personal wellbeing and future aspirations.

Individual interviews were conducted with workers, line supervisors and dormitory managers from the same factories selected for FGDs, as well as with some key industry leaders. These were used primarily to triangulate the information collected through the firm surveys and FGDs.

Data analysis and integration was conducted in a series of phases. After extensive 'cleaning' of the firm survey data, standard statistical analysis was undertaken. These results were compared and contrasted with findings from worker interviews and FGDs. In addition, regression analysis of the determinants of worker departure rates (using firm survey data) was conducted. Finally, a word of caution in interpreting these results is warranted as both the quantitative firm survey data and qualitative data collected through FGDs and individual interviews is largely reporting respondent perceptions. It, therefore, should be interpreted with sensitivity to structural and personal, as well as random, biases of respondents.

Survey results and analysis

Labour supply

The labour market in Lao People's Democratic Republic, although changing, remains highly informal and agriculture based. Factory work, although difficult and demanding, is considered less arduous and offers better income opportunities than alternatives such as family farming or construction work. Most workers see garment work as an interim strategy to build savings in order to start their own small business and/or family.

Prior research had found that the overwhelming majority of garment firms in Lao People's Democratic Republic cite labour-related constraints as their main concern.[9] This finding was largely confirmed in the 2011 survey in which large and medium firm managers overwhelmingly identified 'labour constraints' as their biggest challenge in doing business in Lao People's Democratic Republic (Figure 6.2a); while smaller firms had more difficulties with raising capital and/or securing financing. Across all firm sizes, the majority of managers identified

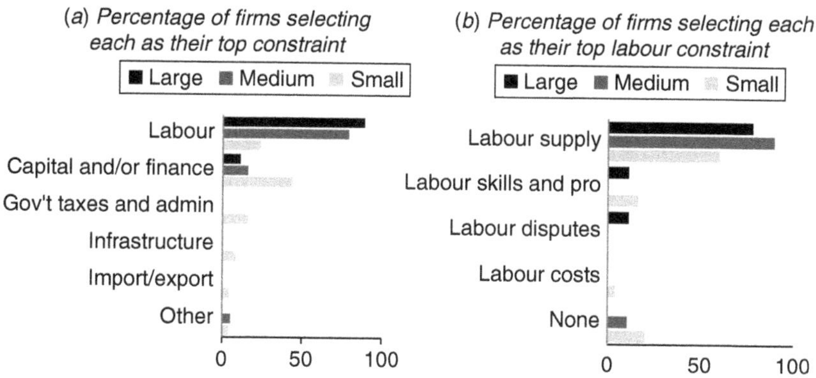

Figure 6.2 Labour is the top constraint for garment firms, and in particular the supply of labour
Source: World Bank Lao garment sector survey (2011).

labour supply as their primary labour-related constraint and labour skills and productivity as their second biggest (Figure 6.2b).

All firms appear to have difficulty attracting and retaining workers with attrition rates (calculated as the proportion of workers who left the factory the year previous to the survey, over the firm's total workforce at that time) from large and medium firms at around 3.5 per cent of their workforce every month, while small firms lose over 6 per cent of theirs. At the same time, large and medium firms report that as many as half of their workforce has been with the factory for three years or more. This suggests that there is a core longer term workforce in the industry, and then a substantial proportion of workers that are constantly entering and exiting. While high attrition rates plague garment industries in many countries, managers of multinational companies operating in Lao People's Democratic Republic report that this rate is high, even for regional industry standards.

Recruitment expectations and experiences

The survey found that firms across all size categories appear to be losing more workers than they are recruiting. Given that managers report having to turn away orders for lack of labour supply, the sector is apparently operating at levels below full capital utilization.

In FGDs, most workers explained that they took jobs in the garment sector for economic reasons: seeking income as a pathway out of poverty, to support siblings' education and to secure a better future for themselves. Many believed factory work to be preferable to working in rice fields or other options in their villages.

> I am working in the garment factory because I do not have a job in my village apart from working very hard in the rice fields for my family.
>
> (Lao garment factory worker)

Most had a friend and/or a relative with experience of garment factory work who alerted them to some of the difficult living and working conditions. Nonetheless, the majority of new recruits were still expecting 'nicer supervisors, more income, less overtime, cleaner and safer working conditions with better accommodation facilities'. But most felt they had little alternative, having neither the skills needed for more attractive 'office' jobs, nor sufficient capital to start a small business.

> For a person who was born to a poor family then ended up with no qualification and no job opportunity, only garment factory or construction worker jobs are available to them.
>
> (Lao garment factory worker)

For most workers, recruitment is swift: many spoke of being hired 'on the spot' without any complex hiring procedures (beyond submitting proof of age) and little explanation of their contractual rights and obligations.

> When I applied for a job here they asked me only a few questions, no introduction of names, no information about work and they did not even tell me about how much I would get for my salary. After they decided to offer jobs [to me and the other new recruits] they called the line manager to bring us to the workplace. There was no time to settle down or prepare - we just started working immediately. We had to leave our personal belongings at the security box until we finished work; later, they took us to the dormitory.
>
> (Lao garment factory worker)

Recruitment procedures are more formalized in some larger factories: workers must be 18 years old or above, able to read and write, and provide proof of age and identification, as well as a letter of guarantee from their village headman. However, most participants in the FGDs reported starting to work in the factory while they were 15 and 16 years old and providing falsified documents. Conditions of employment (factory rules, working hours, salary and overtime requirements) are more clearly explained in some larger factories. Workers complained of the practice in some factories of withholding partial salary of new workers as a 'guarantee' to prevent them from quitting during the first six months of employment; and that these earnings were never repaid. This notwithstanding Lao labour law requirements are to pay 90 per cent of salary during a probation period not to exceed 30 days.[10]

For many employees the reality of factory work is very different from their expectations: hard work and long hours of compulsory overtime, poor living conditions and insufficient income (especially when salary is withheld during the first months of employment) to cover basic living expenses. In particular, the demands of industrial working routines and pressure from managers were completely new and very difficult for many.

> I would say our working place is like a prison because we are not allowed to do anything apart from working for them. We are not allowed to take a break before starting overtime, even though everyone needs to eat something ... We are so hungry sometimes, we keep cookies in our pockets for eating during work, but if they see, we will be fined. You can see we look like malnourished people, we never have enough sleep ... If possible, I would like to have 20–30 minutes break before starting to work overtime in the evening.
>
> (Lao garment factory worker)

Managers report that their employees quit for a number of reasons. Those who leave the industry completely are most likely, according to managers, to do so in order to return home to their communities or due to family responsibilities such as getting married, having children and so on. Workers at larger firms in particular seem more likely to move to other factories within the garment sector, looking for better pay or working conditions. Interestingly, few managers or workers report that workers leave to find jobs in Thailand. This finding is somewhat surprising given the strong 'pull factor' that Thailand is regarded to exert on the Lao labour market.[11] This reflects Lao migration patterns generally, which show migration to Thailand is strongest from bordering southern provinces.

According to several FGD participants, rigid workplace rules and harsh management make the workers lose loyalty and enthusiasm for their jobs; and is a major reason why workers quit.

> When we have a problem in the factory, the only solution is to be patient; if we can't, then we quit. Therefore, when facing this situation, many of the workers who had never worked in garment work before and never had anyone yelling at them like this just ran away.
>
> (Former Lao garment factory worker)

In terms of strategies to improve retention rates, many firms report offering annual bonuses of increasing amounts for numbers of years worked, promises to move into higher paying positions in the production line and various other informal encouragements such as recreation events and celebrations for national holidays. According to workers, managers also make it very difficult for them to resign: withholding partial salary of new recruits and workers' final month salary.

> When someone wants to resign, the factory refuses and puts pressure on us – often allocating that person to the areas that no-one likes such as general section. Some workers could not continue under these conditions and they decided to quit without approval – they just packed their bags and left without any termination pay.
>
> (Lao garment factory worker)

FGD participants seemed aware that garment factories are facing labour shortages and believed that it is easy for experienced workers to move in and out of jobs. They explained that after withdrawing from the workforce for a short period (usually three to six months) workers tend to return to the same factory so they don't have to 'start all over again'.

Labour productivity and training

Research from the recent Investment Climate Assessment for Lao People's Democratic Republic suggests that labour productivity in Lao firms is lower than in most comparator countries and lower than one would expect for a country at this income level. Indeed, productivity in Lao People's Democratic Republic is slightly lower than Cambodia and Viet Nam and significantly lags behind the leading countries in the region such as China, Thailand and Malaysia (Figure 6.3).

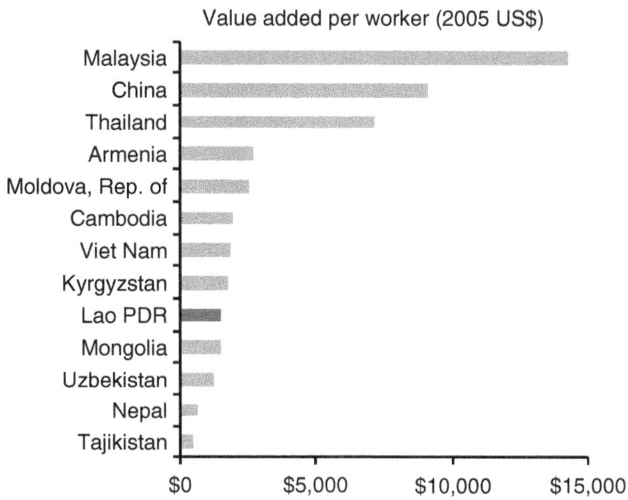

Figure 6.3 Labour productivity in Lao People's Democratic Republic is lower than in the most successful exporting economies
Source: World Bank Enterprise Surveys (2009).

New research also finds that, unlike in most countries, exporting firms in Lao People's Democratic Republic show lower levels of productivity than non-exporters.[12] This finding is counterintuitive as in most countries, exporters are more productive than non-exporters, large firms are more productive than small firms and foreign-owned firms are more productive than domestic firms.[13] This is not the case in Lao People's Democratic Republic. World Bank analysis indicates that although the median foreign-owned firm is slightly more productive than the median domestically-owned firm, non-exporters are more productive than exporters.[14] This appears to be a symptom of 'dutch disease' with large scale investment and rapid growth in the development of Lao natural resources starting to have an effect on the non-natural resource sectors, including artificially raising the returns to investment in the domestic economy (compared to exporting).

When large and medium-sized firms were asked to rate their company's willingness to make various investments in order to improve labour productivity and performance, they expressed strongest interest in the introduction of new production technology and line management techniques and were more sceptical of increasing wages, overtime pay and/or bonuses. However, more than half of managers surveyed expressed some or strong interest in providing more benefits to workers and this jumped to 80 per cent for non-monetary incentives.

All firms provide some basic initial skills training upon recruitment and new hires are quickly put into production lines to continue skills development 'on the job'. When asked to identify priority training needs for their firm's workforce, production skills (cutting, stitching, etc.) and basic literacy and numeracy were identified as priorities for production workers, while production management skills (e.g. line balancing) and quality assurance (control, inspection, etc.) were most commonly identified as top priorities for production supervisors. Large firms also identified team management and motivation as a top priority. Workers in FGDs meanwhile expressed a strong desire for various forms of training and upgrading – particularly in order to be able to earn increased wages in production units. They also expressed interest in literacy and numeracy skills development as well as basic life-skills training, including sexual and reproductive health, and financial literacy – a finding confirmed by recent Participatory Ethnographic Evaluative Research (PEER) undertaken by CARE International to document the risks and opportunities for young female rural migrants to garments factories in Vientiane (Malam 2012).

Although managers identify skills deficits as a major contributor to low productivity levels, and despite the fact that managers had clear ideas about their workforce's training needs, many managers expressed reservations about investing in training for workers. When asked to identify the primary constraint for increasing their firm's investment in skills training, half of survey respondents cited 'loss of return on investment because of high turnover rates' and another quarter cited that 'useful or appropriate training was not available'. Interestingly, only four respondents cited insufficient resources as the primary constraint to greater investment in training.

FGD findings suggest looking not only to workers but also towards management practices to find efficiency gains. In addition to skills deficits, workers suggest that long hours and difficult working conditions act as drags on productivity. Furthermore, line supervisors could point to specific causes of production inefficiency; for example, to bottlenecks in production lines, poor line balancing, last minute orders and rush delivery, and delays in the delivery of inputs. Some of these issues would suggest the possibility of efficiency gains through better production as well as human resource management by Lao garment firms. Other issues relate more to disarticulation between suppliers and buyers mediated through the broader complex dynamics in the global garment supply chain, and largely beyond the remit of individual firms or the Lao garment industry.

Labour relations

Overall, managers painted a fairly positive picture of attitudes and relations in the workplace. Almost 90 per cent of managers surveyed from large firms felt motivation of workers was good, while only half of those from small firms and a quarter from medium firms did. Managers from large firms were also more likely to rate relations between workers and their supervisors or managers as good, than did those from small and medium firms. This is in direct contrast with findings from FGDs in which many workers complained of strict rules and harsh supervision in larger export-oriented factories, while those working in smaller, Lao-owned operations had a better assessment of their working conditions because of 'softer' family-style management.

Participants in FGDs from several factories spoke about harsh behaviour by supervisors and managers as being the most difficult aspect of their working conditions. According to one group, the main reason that workers resign is 'supervisors insulting them and yelling at them when they made mistakes'. Some expressed the feeling of being treated as low-skilled workers, rather than as valued factory employees.

> The most difficult aspect of working in the garment factory is that you have too many bosses and they criticise you in front of everybody and make you feel very bad. Particularly when you are very tired from work and hoping to earn some cash that you could send to your family... it is demoralizing when they shout at you like you are not a human.
>
> (Lao garment factory worker)

There also appear to be cultural clashes between Lao workers and some foreign managers.

> There are many times that foreign staff use impolite and inappropriate words with Lao staff – humiliating us, scolding us and as a result we lose motivation and spirit while the pressure rockets.
>
> (Lao garment factory supervisor)

Worker representation varies by firm size, with nearly all large and medium firms reporting dormitory representatives as well as representatives of the Lao Federation of Trade Unions, while workers in small firms were more likely to have informal representatives. The Lao Women's

Union was also present in just under a third of all firms. However, these representatives have limited independence and former workers complained there was no effective means of conflict resolution on the factory floor as supervisors were fearful of reprisals from managers if they tried to support or defend workers in their production units.

Overall, workers' accounts largely suggest a lack of effective labour representation in the garment sector.

> In our factory there is a trade union officer who is supposed to provide support for workers but he cannot help with anything. In the past I used to report to him about the piece rate having been reduced, and he just told me: never mind, I will talk to them for you. But he never takes action, I guess he is worried they will cut his salary too.
>
> (Lao garment factory worker)

A majority of large and medium firms report receiving inspections by the Lao Ministry of Labour and Social Welfare on labour practices, working conditions, social security and other matters within the ministry's mandate. However, many firms complained that labour regulations and social welfare system requirements were often unclear and arbitrarily enforced. Workers in FGD wanted to know more about labour standards and their enforcement.

> I would like the social welfare department come to check our factory once a month, I want them come to talk to workers in each line directly – to interview workers, not managers, because they [managers] won't tell the truth.
>
> (Lao garment factory worker)

Working and living conditions

Overall, workers assessed their working conditions to be difficult and demanding and, in some cases, complained of being subjected to harsh treatment by supervisors and managers. However, most chose to continue working in the garment sector in order to earn much needed cash to help their families and improve their own prospects. Some expressed hope that their siblings and (future) children would not have to struggle to earn their incomes under similar conditions; and this was in fact a motivation for them to continue working in the garment sector.

Hours

Managers report average working hours per week of between 56.5 (for small firms) and 59.3 (for large firms), including an average of nine hours overtime. This finding contrasts significantly with testimonies from FGDs and individual interviews in which workers complained that overtime is often much longer, poorly paid and often compulsory.

Factories in Lao People's Democratic Republic do not appear to operate in shifts. Regular working hours are reportedly similar across the industry: a regular working day is 8–5 p.m. (with one hour for lunch), six days per week with several daily hours of overtime throughout the week and on weekends, particularly during 'crunch times'. Although Lao Labour Law requires factories to provide every worker with at least one day's rest per week and to seek authorization from the labour ministry (and approval of the workers' representative) for workers to undertake more than 45 hours overtime per month (or three hours on any single day),[15] this does not seem to be an impediment to extending overtime in practice. The Labour Law also clearly provides that workers are meant to be compensated at 150 per cent for regular overtime; 200 per cent for overtime at night (after 10 p.m.); and 250 per cent and 300 per cent for overtime at night, on holidays or during weekly rest days respectively.[16] However, according to workers, these overtime rates are not consistently applied in practice.

Instead, many workers report very low 'fixed overtime rates' that do not reflect Labour Law guidelines. They also complain of extreme exhaustion after being pushed to work 'all night shifts' and punitive measures against those who do not report to work the next morning.

> Overtime work is compulsory and everyone has to attend. If we are absent they will deduct a penalty that is larger than overtime pay from our salary ... If we have to do overtime from 6 p.m. till 6 a.m., we still have to come back to work again the next day at 8 a.m. – even if we did not sleep. If we do not attend work that day, the manager will be very angry – we will receive a warning letter and a fine.
>
> (Lao garment factory worker)

While most workers seem happy to work limited regular overtime to earn extra income, many complain that excessive overtime for little compensation drains their energy and motivation, and is often cited as a key factor in the decision to quit. Those who had a more positive assessment of their working conditions often referred to a 'reasonable amount of overtime' as a key factor.

Wages

Wages for factory workers are calculated differently depending on jobs performed by different segments of the factory's workforce. At the time of the survey, factories reported that base monthly wages varied from US$46 per month for unskilled workers in small factories to US$71 per month for skilled or semi-skilled workers in large factories. However, those working in production lines can earn significantly more (upwards of double these base amounts) and most workers want to be in these units. The salary for most production line workers is calculated based on piece rates and targets set either for individual workers or production units. That is, managers set piece rates and targets depending on the demands of different tasks; workers then get compensated based on meeting and/or exceeding these targets and rates are set based on levels of difficulty and average time spent on each piece. Only a small percentage of small and medium-sized firms (12 per cent and 16 per cent respectively) allow workers to earn additional income by taking extra jobs home.

Workers' views on their wages varied – with those able to achieve higher earnings through the production 'piece rate' system more likely to express satisfaction with their salaries. On the lower end of the salary scale, and for those with dependants, many reported finding it increasingly difficult to cover their basic expenses due to the rising cost of living.

New recruits struggle to understand how piece rate coefficients are calculated depending on the relative difficulty and number of steps to produce an individual garment. Others complained that piece rates seemed to be set arbitrarily and some suspected managers of driving these rates down unfairly.

> Working in the tailoring unit is good because you can get more money if you produce more, but it is very hard work ... and it is not fair because the piece rate is not stable: it is often changed depending on the supervisor.
>
> (Lao garment factory worker)

Accommodation

All large firms, two-thirds of medium firms and half of small firms reported providing on-site accommodation for a proportion of their workers. Overall, 54 per cent of the Lao garment workforce are living in factory dormitories, about one-fifth of whom have small accommodation fees deducted from their wages. About one-third of all firms reported providing other off-site accommodation to workers (especially married workers or male workers who are ineligible for female dorms) and/or accommodation subsidies and allowances. The number of workers reported per dormitory room varied from two to 20, as did the facilities provided in different dormitories.

Migrant workers generally depend on factories to provide them with accommodation in factory dormitories as they cannot afford to pay rent for private accommodation. Many said that they had expected better living conditions when they migrated to work in garment factories, although conditions reportedly vary between factories. Very few workers in focus groups and interviews were satisfied with their living conditions in factory dormitories, complaining in particular of unclean and inadequate water and sanitation facilities, lack of privacy, lack of ventilation and fire safety, insufficient cooking and laundry facilities.

> Some nights I feel very thirsty but there is no drinking water in the dorm so I have to wait till morning to get drinking water at the workplace; those who have more money can buy drinking water ... they treat us as we do not have any feeling and are not human.
>
> (Lao garment factory worker)

Some managers expressed sympathy with the difficult living conditions of workers in factory dorms and made useful suggestions during the survey on how these could be improved, including through inspection and support from local health authorities, expansion of water and electricity supply, increased space for cooking, eating and recreation.

Regression analysis[17]

In this section we analyse the determinants of workforce departure rates data to investigate the linkages between firm characteristics and/or practices with turnover rates as a proxy both for firm productivity (assuming high turnover reduces productivity) and for worker wellbeing (assuming that workers resign to improve their wellbeing). Although we try to account for possible measurement errors and risks of bias due to sample selection, reverse causality and so on, these results need to be interpreted with care given the small sample size (89 firms). Nonetheless, they add some further weight to the qualitative findings from FGDs on the determinants of worker departures.

However, there were a number of key issues raised by workers during FGDs that were not captured in the firm survey data (i.e. not reported by managers) and are, therefore, excluded from this analysis, such as worker accounts of frequent and excessive overtime (beyond what managers reported), firm practices of withholding wages during probationary periods or as fines for minor infractions (contrary to Lao Labour Law provisions and, therefore, under-reported), as well as reports of harsh (sometimes abusive) treatment by supervisors. With these caveats in mind, the regression analysis results suggest that firm size, overtime frequency, working environment conditions, wages and workers' origins are possible explanations for what is described as one of the main constraints to the garments sector performance, that is, high worker turnover rates.

Description of variables

Our variable of interest is the attrition rate at the firm level. It is calculated as the proportion of workers who left the factory in the year previous to the survey, over the firm's total workforce at that time. We regress this variable on a number of independent variables that we suspect to be possible determinants of turnover rates including: firm size, wages, overtime, worker origin (migrant or non-migrant) and working environment.[18]

Regression results

From the ordinary least squares regressions[19] of these variables it appears that the volume of firm sales (corresponding roughly to firm size), the number of overtime work hours and the average minimum wage for both skilled and unskilled workers are the most important determinants of attrition rates. Indeed they remain statistically significant and with expected signs to the inclusion of other controls such as the working environment proxy or workers' origins. Size dummies turn out to have a positive and significant impact only when workers' origins are included. Interestingly, while a higher minimum wage for unskilled workers leads to a substantial decrease in attrition (a 1 per cent increase in unskilled minimum wage reduces attrition by more than 1 per cent in all specifications) as would be expected, we find the reverse effect for the wage of skilled workers. Using a summary index or a set of dummies for the working environment did not show statistically significant impact. Overtime work hours are the most robust determinant across specifications – as expected from the qualitative survey – though with a small magnitude: an additional overtime work hour suggests an increase in the firm attrition rate by 0.044–0.115 per cent.[20]

Though such a regression analysis is subject to a wide range of biases and caveats that cannot be totally ruled out (small sample size, reverse causality, omitted variables), we can nonetheless gain increased confidence in the finding from worker interviews that, in addition to difficult working conditions and relations between workers and supervisors (which were not adequately captured in firm survey data), high overtime frequency and low levels of minimum wage for unskilled workers may be the most important determinants of the observed high turnover rates in the Lao garment sector.

Conclusions

This chapter began by exploring the reasons for high turnover and lower productivity in the Lao garment workforce. By undertaking both a structured survey with managers and FGDs with workers, the sharp differences in their perspectives were revealed particularly on questions of workplace relations and conditions. Taken together, we conclude that reasons for workforce departures are multiple. They include, namely: difficult transitions (especially for rural migrants with insufficient information on working and living conditions or understanding of their contractual rights and obligations); wages (both actual rates and a perceived lack of consistency and transparency); and working hours (particularly excessive overtime demands). These were identified as the primary reasons workers quit. Interviews with workers also revealed that they resent pressure and conflicts with supervisors, and that these can be further exacerbated by cultural clashes with foreign (usually male) supervisors. The lack of effective systems of dispute resolution or worker representation provides limited means of resolving problems on the factory floor. Many workers feel they simply have to 'put up' with these problems or vote with their feet by quitting; thus, further exacerbating the industry's problems of high turnover and low productivity.

These findings suggest some avenues for improvement in both labour standards and productivity in the Lao garment industry, and that indeed these objectives are, at least to some extent, compatible. Many of these changes would focus on improvements in working and living conditions, strengthening worker representation and dispute resolution mechanisms within factories and across the industry, as well as on modernization of human resource and production management systems. These efforts could be significantly enhanced by longer-term investment in building a skilled and healthy workforce through enhanced public–private partnerships and programmes.

The Lao garment sector has also been struggling with access to premium buyers due in part to a perceived lack of credible information on the industry's labour standards and management practices. This study points to possibilities for increasing worker wellbeing, reducing turnover and improving productivity through mutually reinforcing actions. Possible interventions include: a transition support programme and improved labour market information systems to support labour force entry; better worker representation and dispute resolution mechanisms; and a factories improvement programme that would include monitoring and certification systems to provide incentives for firms to improve conditions as a means of accessing elusive premium buyers. While recognizing that margins remain tight and opportunities for major investments or wage increases are limited by the realities of a competitive global garment market – in which Lao firms' weak position in global value chains leaves them mostly as price takers (Staritz 2010) – we believe that changes in policies and practices at the level of firms, industry and public sector could have positive results for Lao garment workers and industry.[21]

There were a number of limitations to this study, some of which suggest possible lines of further inquiry. One interesting area for further analysis would be to attempt to find any linkages between the characteristics of firms, their policies and practices (such as firm participation in social compliance schemes or working and living conditions in factories), with outcomes in terms of productivity and/or labour standards. This would probably require further survey data collection and the development of composite indices (or clever proxies) for measuring labour standards outcomes.

The lack of systematically collected or reliable labour market data in Lao People's Democratic Republic did not allow for comparison of workforce entry and departure rates across sectors. While donor-supported efforts are underway to strengthen the government's capacity in this regard, the current dearth limits contextualization of garment sector labour trends within the Lao labour market or South-East Asian regional market (including comparisons with other industries or sectors).

This study is also limited by the fact that data collection from workers was primarily through FGDs. In the future, complementing these rich qualitative insights with quantitative data collected through worker surveys would strengthen the capacity to present a representative picture of workers' experiences and perceptions. It would also help in the analysis of, and comparison between, particular subgroups of workers. That is, to be able to better understand the experiences and differences between, for example, male and female workers, long-term and short-term employees, migrant and non-migrant workers, and so on. The collection of panel data in particular would allow for greater analysis of trends in this labour market and possibly even of the impacts of key policies or programmes to improve productivity and labour standards, which may be introduced in the future.

Finally, mapping of the Lao industry's position in the global garments value chain and analysing the constraints this places on managers and workers (which in turn shape their perceptions) would be an interesting avenue of further inquiry and contribution to our understanding of how local labour and productivity challenges are mediated by global value chains.

Annex: Regression results

Model 1 Determinants of attrition rates

Dependent variable: log attrition rate								
OLS	**(1)**	**(2)**	**(3)**	**(4)**	**(5)**	**(6)**	**(7)**	**(8)**
Firm size	–0.309 (0.165)			0.032 (0.263)				
Medium		**–0.517*** (0.271)	–0.266 (0.316)		–0.179 (0.314)	0.368 (0.325)	0.424 (0.363)	**0.738*** (0.419)
Large		–0.508 (0.350)	0.147 (0.559)		0.294 (0.555)	0.904 (0.559)	0.943 (0.577)	**1.350**** (0.654)
Ln sales (LAK)			–0.113 (0.076)	**–0.129*** (0.076)	**–0.145*** (0.077)	**–0.173**** (0.078)	**–0.168**** (0.081)	**–0.162*** (0.090)
Overtime work				0.044* (0.023)	0.043* (0.026)	**0.084***** (0.030)	**0.085***** (0.030)	**0.115***** (0.036)
Ln unskilled wage						**-1.377**** (0.599)	**-1.385**** (0.607)	**-1.594**** (0.622)
Ln skilled wage						**0.746*** (0.425)	**0.739*** (0.431)	**0.833*** (0.436)
Work environment							–0.034 (0.094)	–0.077 (0.098)
Lao migrant (%)								–0.007 (0.005)
Foreign migrant (%)								–0.064 (0.102)
Work environment dummies	No	No	No	No	No	No	No	No
N	50	50	50	50	50	41	41	41
R^2	0.06	0.08	0.09	0.15	0.18	0.32	0.32	0.37

Note: p<0.10*, p<0.05**, p<0.01*** Estimates for log variables should be taken as elasticities while others give the attrition rate returns to scale.

Model 2 Instrumenting firm size by age

Dependent variable: log attrition rate				
2SLS	**(IV 1)**	**(IV 2)**	**(IV 3)**	**(IV 4)**
Firm size	–0.913 (1.399)	–0.878 (1.421)	0.278 (1.387)	0.155 (1.065)
Ln sales (LAK)	0.098 (0.317)	0.081 (0.331)	–0.135 (0.312)	0.078 (0.193)
Overtime work		0.029 (0.038)	**0.082**** (0.039)	**0.091*** (0.055)
Ln unskilled wage			–1.318 (0.955)	–1.272 (0.853)
Ln skilled wage			0.703 (0.699)	0.655 (0.540)
Work environment				–0.018 (0.136)
Lao migrant (%)				–0.003 (0.010)
Foreign migrant (%)				–0.055 (0.098)
Work environment dummies	No	No	No	No
N	50	50	41	41
R^2	0.06	0.08	0.31	0.18

Note: p<0.10*, p<0.05**, p<0.01***.

Model 3 Looking at the deviation of attrition rates from the mean

Dependent variable: deviation of attrition rates from the mean								
OLS	**(1)**	**(2)**	**(3)**	**(4)**	**(5)**	**(6)**	**(7)**	**(8)**
Firm size	**–0.189**** (0.090)			–0.104 (0.149)				
Medium		**–0.323**** (0.147)	–0.244 (0.175)		–0.228 (0.178)	0.094 (0.148)	0.108 (0.165)	0.256 (0.190)
Large		–0.307 (0.190)	–0.101 (0.309)		–0.074 (0.315)	0.302 (0.255)	0.312 (0.263)	0.524 (0.297)
Ln Sales (LAK)			–0.036 (0.042)	–0.032 (0.043)	–0.041 (0.044)	**–0.066*** (0.117)	**–0.065*** (0.037)	**–0.067*** (0.040)

(*continued*)

Model 3 Continued

Dependent variable: deviation of attrition rates from the mean								
OLS	(1)	(2)	(3)	(4)	(5)	(6)	(7)	(8)
Overtime work				0.008 (0.015)	0.008 (0.032)	**0.037***** (0.013)	**0.038***** (0.014)	**0.052***** (0.016)
Ln unskilled wage						**–0.663**** (0.273)	**–0.665**** (0.277)	**–0.766**** (0.282)
Ln skilled wage						**0.383*** (0.193)	**0.381*** (0.196)	**0.420**** (0.198)
Work environment							–0.008 (0.043)	–0.028 (0.044)
Lao migrant (%)								–0.004 (0.002)
Foreign migrant (%)								–0.019 (0.046)
Work environment dummies	No	No	No	No	No	No	No	No
N	50	50	50	50	50	41	41	41
R^2	0.08	0.11	0.12	0.10	0.13	0.17	0.32	0.37

Note: p<0.10*, p<0.05**, p<0.01***.

Notes

1. This chapter is based on World Bank Report No. 70809-LA: 'Lao PDR: Labour Standards and Productivity in the Garments Export Sector – A survey of managers and workers' (July 2012). This research was funded by the World Bank's Gender Action Plan and by the Lao PDR Trade Development Facility, a multi-donor trust fund financed by Australia, the EU and Germany, and administered by the World Bank. An earlier version of this chapter was presented and helpful feedback received at the ILO/IFC Better Work Conference in Washington, DC, on 26 October 2011.
2. The findings and interpretations expressed here are those of the authors and do not necessarily reflect the views of the World Bank, its executive directors or the countries they represent.
3. Phouxay and Tollefsen (2011) is a notable exception.
4. A 2006 study by the Lao National Statistics Centre focused on three main aspects of competitiveness in the Lao garment sector: labour costs, labour and capital productivity, and lead times. The study found that the hourly wage in Lao People's Democratic Republic was considerably lower than in China (US$0.12 versus US$0.68) but that Lao labour productivity in the sector was also substantially lower (NSC 2006).

5. The US market has grown in importance since the establishment of Normal Trade Relations in 2005, but still remains of limited importance to the Lao garment sector.
6. Labour Law (Amended), Passed by National Assembly 27 December 2006; Promulgated by Presidential decree No. 05/PO, 16 January 2007.
7. Lao PDR has ratified five out of the eight ILO Conventions corresponding to core labour standards, and the revised Labour Law of 2006 includes key principles from these ratified conventions. Lao People's Democratic Republic has ratified the following ILO Conventions: the Forced Labour Convention, 1930 (No. 29); the Equal Remuneration Convention, 1951 (No. 100); the Discrimination (Employment and Occupation) Convention, 1958 (No. 111); the Minimum Age Convention, 1973 (No. 138); and the Worst Forms of Child Labour Convention, 1999 (No. 182). Although Lao People's Democratic Republic did not ratify the Freedom of Association and Protection of the Right to Organise Convention, 1948 (No. 87) or the Right to Organise and Collective Bargaining Convention, 1949 (No. 98), elements of these core labour standards are included in the 2006 Labour Law (although within the socialist unitary state model of a national workers' trade union). The only Convention that has neither been ratified nor reflected in conforming legislation is the Abolition of Forced Labour Convention, 1957 (No. 105).
8. This scale of garment firm size was developed specifically for this study of the Lao garments sector to reflect observed segmentation of the industry, which in turn allows for analysis of relationships between specific firm characteristics/profiles and labour practices and/or productivity outcomes. This scale differs from the Lao Government's categorization of firms as micro/small/medium/large as those having respectively, 1–4, 5–19, 20–99, 100+ employees. According to this classification almost all Lao garments firms would be considered either medium or large.
9. World Bank (2011) op. cit.
10. Lao Labour Law, op. cit., article 27.
11. A possible explanation may be that migrants choose between Thailand and the Lao garment sector as initial alternatives when seeking employment, rather than switching between the two. Certainly, the Lao garment sector attracts fewer migrants from the southern provinces of Lao People's Democratic Republic (bordering Thailand) than from northern provinces.
12. World Bank (2011) op. cit.
13. Note that size categories (small, medium, large) used in the Enterprise Survey (which includes firms from all sectors) differ somewhat from those used in this Garments Sector Survey. However, the relationship between firm size and labour productivity across the larger formal private sector found in the Enterprise Surveys appears to also hold in the garments sector.
14. This is a result taken from the 2009 Lao Enterprise Survey (and discussed in the 2011 Investment Climate Assessment), and holds for all manufacturing firms in the Enterprise Survey sample, of which garments manufacturing represents around a third. It is a counterintuitive result, but likely driven by the emerging signs of Dutch disease associated with the Lao natural resources boom, as well as low competition in the domestic market. This is driving up the returns to non-exporters, while exporters struggle with a costly supply chain and low productivity, which erodes competitiveness.

15. Lao Labour Law, op. cit., articles 18 and 19.
16. Lao Labour Law, op. cit., article 48.
17. A full description of the regression analysis summarized here can be found in Annex B of the full report (World Bank 2012).
18. For working environment we include a set of dummy variables indicating whether the firm provides various social security benefits, transportation, meals, on-site and/or off-site accommodation, and whether the firm implements a human resources plan. As an alternative, and because our number of observations is very small, we build up a simple summary indicator ranging from zero to one in which all mentioned variables enter equally. Note, however, that this does not allow for any consideration of the quality (actual or perceived) of any particular benefit; merely whether it is provided or not.
19. Note that we cannot use a Probit regression model in that setting: even though our dependent variable is a ratio, it is not necessarily bound between zero and one. Indeed, some firms (in particular small ones) have attrition rates larger than 100 per cent in one year. The total number of workers leaving and entering the factory over a year is greater than the number of workers employed at any one time. Hence our use of a linear OLS regression model instead.
20. Note OLS estimation does not prevent the risk of reverse causality in the model. High attrition rates may impact a firm's size, economic performance or need for OT work hours. In order to isolate the causal impact of our dependent variable on attrition rates, we need to find a set of robust instruments for all possible determinants. Unfortunately, both sample size and the limited amount of information available prevent us from identifying and using good instruments. Looking at the covariance matrix between our dependent variables and other firm characteristics, such as its ownership structure, percentage of output to export and so on, shows almost no statistically significant relationship. The only significant correlation (at the 10 per cent level) we find relates positively a firm's size and its number of years of operations in Lao People's Democratic Republic. Furthermore, it is unlikely that a firm's 'age' may directly impact attrition rates. We, therefore, attempt to instrument firm size by age in a 2SLS model. Using such an instrument, we should be aware that we mechanically increase the imprecision of our estimates – interpreting the results with such a small sample is then an even harder task. Thus, unsurprisingly most of our estimates lose their significance and display large standard errors. The only factor that seems partially robust to this 2SLS strategy is the number of overtime work hours: in the two last columns an extra OT hour increases the attrition ratio by 0.081–0.091 per cent, results in the range of those found previously.
21. For a fuller discussion of policy implications, see section 9 of the full report (World Bank, 2012).

7

Vulnerable Workers and Labour Standards (Non-)Compliance in Global Production Networks: Home-Based Child Labour in Delhi's Garment Sector

Resmi Bhaskaran, Dev Nathan, Nicola Phillips and Upendranadh Choragudi[1]

Introduction

The South Asia region retains the highest incidence worldwide of child labour. In India, figures from the 2007–2008 Labour Force Survey indicate that working children account for around 2.4 per cent of the total population aged between 5 and 14 years (IAMR 2011),[2] which ostensibly contrasts favourably with much higher figures of 42 per cent for Nepal or 19 per cent for Bangladesh (Chakrabarty and Grote 2009) and 6.2 per cent in India itself in 1993–1994. But in terms of absolute numbers of working children, India has the highest concentration in the world, with estimates indicating a total of 12.6 million. In some states the incidence of child labour exceeds 6 per cent (Rajasthan, Gujarat, Orissa, Chhattisgarh and Andhra Pradesh), and it is higher in rural areas than urban areas. Child labour persists across many sectors, including those such as the garments industry, which are tightly integrated into global production networks (GPNs) (Posthuma and Nathan 2010).[3]

This chapter explores the nature of child labour in segments of global garments production located in the National Capital Region (NCR) of India, and reflects on the core governance and policy challenges associated with addressing the problem. In the late 1990s it was estimated that Delhi accounted for somewhere around a third of national garments production (registered and unregistered production combined), and that it contained some 18 per cent of garments enterprises in India (see Barrientos et al. 2010: 132). In this sectoral context, there has been a reduction in the number of child workers of about 2 per cent per annum between 1999 and 2009 (Nathan et al. 2013), as a result of a concerted effort to remove child labour from the factory setting in response to growing public and political pressure within India and beyond. However, we show that in the process the problem has not been eliminated but instead partially displaced to the arena of household activity, as the restructuring of garments production to reduce costs and evade regulation has dramatically increased the importance of home-based work and the numbers of (particularly women) workers in the burgeoning household sector. Our study also reveals that child labour exists almost entirely in the embroidery and embellishment tasks of garment production, which has come to be concentrated heavily in the household sector.

The first two parts of the chapter briefly set out the methodology of the study and the profile of child labour in the Delhi garments sector on the basis of its key findings. The third discusses issues of governance and social compliance in this sector, identifying the manner in which the organization of production processes and the shortcomings of regulation systems have actively enabled the persistence of child labour. In the fourth part of the chapter we turn to a consideration of the associated policy issues and possible avenues towards the elimination of child labour, arguing that these need to combine initiatives relating to the organization and governance of production networks with various forms of active social policy. We draw particular attention to issues of regulating working conditions and addressing wages in production networks, the significance of the Right to Free and Compulsory Education Act, passed in 2009, and the role of conditional cash transfer (CCT) schemes.

Methodology

The study was based on field research conducted from late 2009 to mid-2010 in New Delhi and the NCR of India. Supplementing basic secondary data from the National Sample Survey (NSS), a household survey in five key locations – Nangloi, Gandhinagar, Seemapuri, Jehangirpuri and Shahpur Jat – collected information on the incidence, profile and patterns of child labour, and the socio-economic characteristics and conditions of the households in which it occurred. The locations selected for the survey are poor, 'semi-slum' areas with a high concentration of garment workers and household-based enterprises in the garment industry, situated near major garment-producing centres or wholesale garment markets. A sample of 220 households was identified across the five locations. Nineteen of these 220 were discovered to be household enterprises – that is, workplaces-cum-hostels where employees work and live along with their employers. These are treated separately in our analysis from 'home-based' households, where the workers are family members. In the 201 'home-based' households, there was a total of 552 children, 370 of whom were in the age group of 5–14 years. The sample is thus indicative rather than representative, but nevertheless sufficiently sizeable to generate revealing insight into the incidence and nature of child labour in the Delhi garments sector.

The second element of field research consisted of 30 firm-level case studies based on interviews. The aim was not to trace direct connections from the sample households to specific firms, but rather to obtain a broad understanding of production operations, firms' engagement with global brands and the domestic market, outsourcing and subcontracting practices, and labour management strategies. The firms and enterprises were selected for their locations in the major garment-producing centres of Gurgaon and the New Okhla Industrial Development Authority, respectively located on the borders of Delhi in the states of Haryana and Uttar Pradesh, and served by activity in the locations included in the household survey. The firms were of diverse types, including main producers/factories, subcontracting firms, intermediary firms, wholesale dealers and designers. The informants were owners, managers and merchandisers. These interviews were supplemented by more informal conversations with representatives from civil society organizations and auditors involved with social compliance issues.

Undertaking research on child labour has always been demanding and challenging (see Mehta and Sherry 2009; Sharma et al. 2004), but has become more so given the spread of awareness of the laws regarding child labour and the stigma attached to this practice. In the areas selected for the household surveys, especially Shahpur Jat (which featured in 2007 in a major exposé of child labour in global production) and Gandhinagar, the presence of strangers, especially those who appear educated and middle class, is frequently met with hostility and obstruction. Equally, it is difficult to speak to the children, as they are always under strict supervision and tutored in how to respond to questions about their age or connections with a household. Such difficulties were addressed in our study by using the services of local facilitators, some of whom were also the enumerators who conducted the survey, and personal connections in the areas. The identity of households and firms has been kept confidential and all conversations were conducted on the condition of anonymity.

Child labour is defined in our study as involving workers between the ages of 5 and 14, in which range child labour is ostensibly prohibited under Indian law. We also define child labour as children's involvement as workers in productive activity within the garments sector, which may or may not be conducted in tandem with schooling, and for which the child may or may not receive direct payment. Distinctions between (paid) child labour and (unpaid) child work, which have been the source of much debate in recent years (see Burra 2005; Leiten 2002), are both problematic in general and inapplicable in the more specific context of work within GPNs. The children who constituted the focus of the present study were often unpaid workers within the context of a family or home-based unit, their contribution being subsumed into a payment made to the family or to the adult workers. Yet this does not negate their participation as workers in productive activity, or the value that accrues within the production network from their labour. The key point is that the commercial dynamics of GPNs function to favour and reinforce the use of this kind of labour, both directly (by fostering the integration of children as workers, whether paid or unpaid, in productive activity) and indirectly (by favouring the use of children in the reproductive household economy so that adult workers are made available for wage labour).

The profile of child labour

The household survey revealed a number of key characteristics of child workers and their households, which can be summarized as follows:

- Sixty-nine per cent of households surveyed had child workers, and 74 per cent of all children in surveyed households worked.
- Children in the age group 5–11 years worked an average of three hours per day; while children in the age group 12–14 years worked an average of four and a half hours per day.
- School attendance was quite high at 84 per cent for the 5–11 group, but it fell to 60 per cent for those in the 12–14 group.
- A high proportion of child workers, 45 per cent, did not receive any specific payment for their labour. For those that did receive payment, the average monthly earnings were about Rs.400 for children aged 5–11 and about Rs.550 for children of 12–14 years of age.
- For almost all child workers (92 per cent), the workplace was the home, and a handful of children worked in other houses or workshops in the neighbourhood.

- Children were largely helpers or workers in embroidery and embellishment work, with older children also engaged in activities such as thread cutting. Ornamentation was the focus of work of households with child labour.
- The major reason given for children to work was to supplement family income (61 per cent of households gave this reason and nearly 85 per cent of children fell into this category), even though, significantly, child workers' average contribution to household income was just 10 per cent.
- A high percentage (26 per cent) of child workers were not engaged in garments production, but rather had dropped out of school in order to look after younger siblings.
- The average per capita monthly income of households with child labour was less than half that of households without child labour. In daily terms it amounted to about Rs.60 per day for households with child labour, clearly below the per capita poverty level of US$2 per day or approximately Rs.85 per day. For households without child labour the average per capita daily income was about Rs.120 per day, which is above the poverty level.
- There is a clear gender dimension to the incidence of child labour, with female-headed households having a higher incidence of child labour than male-headed households. Similarly, the per capita daily income of all female-headed households was just above Rs.60 per day, while that of male-headed households was Rs.133 per day, showing that female-headed households are both poorer and, consequently, more prone to have working children.

Governance, social compliance and child labour

The research revealed that embroidery and embellishment are the main tasks in which child labour is involved in the Delhi garments sector, reflecting a complex system of outsourcing and subcontracting from the major exporting units to the household level. Embroidery and embellishment are the primary functions outsourced by the major exporting firms through subcontractors to the household sector. Some other tasks, such as thread cutting, are also regularly outsourced to the non-factory sector, including the household sector. Significantly, while the value of outsourced work may be just 5–10 per cent of total value of the garments, embroidery and embellishment work is crucial to Delhi's unique niche in garment export.

The use of child labour is so widespread in the household segment of garments production precisely because it is possible thereby to avoid regulation. Child labour is permitted under Indian law at the household level: Clause 3 of the Child Labour (Prohibition and Regulation) Act of 1986 prohibits the employment of children in certain occupations and processes with the proviso that 'nothing in this section shall apply to any workshop wherein any process is carried on by the occupier with the aid of his family'. This exclusion has been pivotal in displacing child labour from the factory to the household sector, resulting from the reconfiguration of outsourcing strategies on this basis. Even in household units, where the workers tend to be young migrant boys from rural areas, it is consequently always claimed that they are relatives of the employer in order to comply with the legislation.

Thus, in terms of the typology put forward by Ravi Kanbur (2009), the displacement from factory to household sector means that child labour law either becomes non-applicable or is difficult to enforce.[4] Yet the difficulties of regulation are related not only to legislative shortcomings but also to a broader set of issues concerning governance and compliance in the garments industry. The key complexity relates to the structures and strategies of outsourcing, which create extremely high levels of market segmentation and fragmentation. Put simply, the longer the outsourcing or sub-contracting chain, the more difficult the governance of the industry becomes, whether with regard to technical quality or compliance with social norms.

Quality norms are usually enforced through rejections, which can vary from 15 per cent to as much as 30 per cent and works right down the outsourcing or sub-contracting chain. For export garments there are clear norms for dyeing and printing, including on the use of chemicals. For exporting firms the incentives for compliance with quality standards are obviously extremely high. Conversely, not much attention tends to be paid to compliance with labour norms, other than those relating to child labour. Audits are performed, but the timing of these audits is usually known well in advance and so it is possible to hide many violations. As pointed out by Barrientos et al. (2010), factory audits are reasonably effective in matters that can be verified in a straightforward visual sense, such as provision of toilets, first aid, absence of child labour on the factory floor and so on. But compliance on payment of overtime and other labour conditions is more difficult to establish and the mechanisms of auditing are much less effective. Further, insiders in the industry told us in interviews, under conditions of anonymity, that it is not difficult to purchase required audit clearances. If factory inspectors can be bought off, there is no reason to presume that other auditors cannot be. This, according to insiders, does not invariably happen, but does occur on a fairly noticeable scale, with a price of around US$2,000 per audit.

The major social compliance issue, however, is that of child labour. Any scandal surrounding the use of child labour has a substantial impact on brand image, in turn affecting market share and shareholder value. As a consequence of the much greater recognition of the business case for eliminating child labour, action has been taken to the extent that the key factory areas can now legitimately claim to be 'child labour free' zones. This represents a clear advance from the 1990s when there was no such concern about the use of child labour.

However, the picture changes when we move to the non-factory sector. In the numerous unregistered small units in housing colonies in the urban village[5] localities, with production largely serving both the low-end domestic market and the export market, there is a visible presence of child workers. The many layers of outsourcing that characterize garments production enable employers and producers to disclaim obligations to workers, and specifically to evade responsibility for child labour. As one small-scale producer put it very simply in our interviews, 'when we give work, we do not bother to check who does what'. As Balwant Singh Mehta and Karren Sherry (2009: 653) also point out in their study of child labour in *zardosi* (a form of embroidery) in and around Lucknow, '[i]n these situations, heads of households are paid on behalf of the entire family, thus allowing employers to disclaim knowledge regarding who is actually working for them, children or adults'. The type of contract between the exporter and outsourced unit is also critical. These range from a full written contract, containing conditions referring to quality norms or the non-involvement of child labour, to purely verbal 'agreements', which are prevalent when working with unregistered units, including household and home-based workers. Such arrangements further enable employers to manipulate working conditions (including pay), evade responsibility to workers and escape forms of regulation such as auditing and labour inspections. Labour inspection systems are in any case often partial and sometimes actively hindered by state authorities (see Deshingkar 2009; ILO 2006). Inspections within the household sector are also particularly difficult, given the inaccessibility of the semi-slum areas in question, and attempts tend immediately to be obstructed by residents and employers in the area.

It thus seems to be the case that the proliferation of levels and layers of intermediary involvement in the completion of the work and, in particular, the involvement of household-based work promote the negation of quality norms and weaken the technical and social quality checking process. Low-skilled production activities that are priced at very low levels are seemingly more prone to have a high incidence of violation of socio-ethical norms, especially the involvement of child labour. Activities that require specified skill sets, on the other hand, lead to a reduction in the number of children in the work force. However, the garment value chain relies on a wide range of unskilled activities and also features a prevalence of small units offering apprenticeship, normally without any payment.

Yet, alongside the role of intermediaries in the organization of production, the other key mechanism by which regulatory structures are undermined relates to labour recruitment. The use of labour contractors is recognized as increasingly prevalent in GPNs in general (Barrientos 2013), and in the Indian garments sector as a whole it is estimated that some 40–50 per cent of the total work force is recruited by labour contractors (*thekedaars*). The consequences for labour standards, including in terms of the forms of child labour we are discussing here, are various. Workers recruited by contractors are usually tied to a particular employer and brought for a specific job. For migrant workers particularly, their possibilities of opting out of the circulation loop, by changing employers or settling in the destination, are thereby severely constrained (Breman 2010: 4). Workers are sometimes employed directly by the labour contractor rather than the firm for which they work, such that firms are able to evade not only obligations to the workers, but also the requirements of social compliance imposed by transnational lead firms or first- and second-tier suppliers. The contractor thus functions both as intermediary and point of rupture in the employment relationship, further distancing the lead outsourcing firms from the workers involved in the production process. All of these wider problems take on particular force given the additional vulnerabilities of migrant child workers, who are most often brought from rural villages on the basis of advances on their wages paid by contractors to their parents, and for whom the workplace also functions as their living quarters.

We can hypothesize that production networks have purposefully been organized and reorganized in order to move a large part of economic activity beyond the reach of governance and regulation. Government policy and legal frameworks have been tailored in ways that support those processes, particularly in the deregulation of the private sector and mechanisms to impose greater discipline on labour in that context. The conjunction of the limitations of private and public regulation has thus put in place a system characterized by what Posthuma (2010) calls 'regulatory enclaves', which exclude the vast majority of workers in the Delhi garments sector, and essentially all child workers. Public authorities are also poorly equipped, or perhaps politically unwilling, to deal with the 'mobile poor' (Breman 2010; Mosse 2010: 1161) – and, it should be added, the increasingly complex and ubiquitous system of labour contractors that organizes and controls them.

Our key point is that child labour needs to be understood not simply (as it often is) as a kind of cultural remnant or the consequence of poverty, but rather also as the *direct* result of the ways in which contemporary production networks are organized and governed. Arguments that see economic development as the means to ending child labour are, therefore, likely to be misplaced unless they take this into account. It is of course true that child labour has deep roots in poverty and deprivation, as this research clearly revealed, but it does not then follow that the solutions lie in economic growth or economic upgrading alone. Indeed, it has been widely acknowledged that child labour does not correlate in any straightforward or predictable sense to patterns of growth, levels of gross domestic product (GDP), or the effectiveness or otherwise of anti-poverty programmes based on income measures (Bachman 2000; Chakrabarty and Grote 2009; Degraff and Levison 2009). There are several reasons for this. First, there is no necessary correspondence between the pace of growth and the pace of poverty reduction. Second, child labour is strongly associated with particular kinds of social norms, which sanction particular kinds of inequality and exploitation, relating especially to gender, caste, ethnicity, race and so on (Mosse 2010; Phillips et al. 2011; Tilly 1998). Third, the persistence of child labour is deeply rooted in the organization and functioning of contemporary economies and production networks, including those involving global lead firms. For this reason the issue of policy strategies is far from straightforward, and it is to this issue that we now turn.

Policy, regulation and governance strategies

Arguments are frequently put forward that not all forms of child labour are unacceptable, and that a strategy of eradication is, therefore, misplaced (e.g. Fallon and Tzannatos 1998; Leiten 2002). The key idea in this sort of argument is that child labour functions as a form of apprenticeship, which is to the benefit of both the child worker and the household. It is also claimed that child labour does not necessarily jeopardize the future earning potential of the child (e.g. Ghose 2004). We reject these claims not only on the grounds that the persistence of child labour in India is clearly at odds with both global and ILO norms and new norms established in Indian law, but also because our research reveals a clear function for child labour in the intergenerational transmission of poverty.

At a basic level, our survey revealed that patterns of child labour in Delhi garments production is substantially at odds with ILO guidelines. As indicated already, the extent of work performed by the children in the sampled households and, therefore, its economic and social significance, are considerable – equivalent on average to some three hours per day for 5–11 year-old children, and four and a half hours per day for those in the 12–14 age group. These figures nevertheless conceal variation. In the 5–11 age group, the survey indicated that more than half those sampled work only for one to two hours per day, 26 per cent work for three to four hours and nearly 16 per cent for more than seven hours. This latter group thus classifies as full-time workers, even though they may well also go to school. In the 12–14 age group, the majority of the children worked for more than three hours a day, and over 17 per cent for more than eight hours. These patterns are in violation of norms stipulated by the ILO[6] of no work at all for children aged 5–11 and 'light work ... not detrimental to education' of not more than two hours per day for children aged 12–14 (see Diallo et al. 2010: 21).

India is also a signatory to the Convention on the Rights of the Child, which stipulates that 'children's work should not jeopardize any of their other rights, including the right to education, or the right to relaxation and play' (Article 32), and in 2009 passed a Right to Free and Compulsory Education Act, making access to education a fundamental right for children. Yet the authors' research reveals that patterns of child labour are at variance with both those provisions, inasmuch as child labour unambiguously corresponds with the absence or premature suspension of education. In India since the 1980s, school attendance has improved significantly and literacy rates have increased, even though overall levels of education remain low, especially among the poor and among women, and real expenditure on education as a proportion of GDP has steadily declined (Drèze and Sen 2002). Child labour, while still prevalent, more often than not occurs in combination with some schooling, at least for younger children who are less valuable in labour markets. Yet, in this survey, 20 per cent of children aged 12–14 were 'drop-outs' from school, rising to 42 per cent in the 15–17 age group, suggesting that child labour at an early age is strongly connected with premature abandonment of education. The illiteracy rate among working children was also nearly double the level for all of the children in the sample. This finding runs counter to the common argument noted earlier that child labour during off-school time does not jeopardize the future earning capacity of the child worker.

The picture also has an important gender dimension. On the one hand, in the 5–11 age group a greater percentage of girls than boys dropped out after enrolling and attended school less regularly. In the 12–14 age group more boys were reported never to have enrolled in school, which is likely to be due to the large numbers of migrant male child workers. On the other hand, as we have seen, gender differences emerge clearly in the reasons for which children do not attend school, as girls are more likely to leave school to look after younger siblings and support the family by working, and older girls are additionally likely to leave school for 'safety' reasons. This restriction of the mobility of girls and women is an important mechanism by which forms of poverty and vulnerability to exploitation are entrenched and socially sanctioned. At the same time, older boys are expected to shoulder economic responsibility and hence are likely to opt for work rather than school, such that the low level of educational attainment is characteristic across the population of child workers.

Thus, neither the apprenticeship nor the part-time, after school-hours nature of child work can be used to justify the continuance of child labour. In various ways, there is a clearly deleterious effect on the capabilities of the children as a result of their dropping out of school. The apprenticeship argument also falls down in the sense that the skills of embroidery and embellishment work are easily acquired, which is why the return to these forms of labour is so low, and that, in the modern world, acquiring a general education through high school is of greater value than the 'learning by doing' skill of embroidery and embellishment. Thus, child labour in the garment industry, by increasing the rate of school drop-outs, locks the children's future into the lowest-earning rungs of garment work or any other similar activities. The very clear correspondence in our survey between low overall levels of household education and the incidence of child labour further reveals the intergenerational transmission of poverty on this basis.

As Indian law[7] itself now calls for the elimination of child labour that affects the educational attainment of the children, the basis for further action to address the problem is clear. Yet the appropriate strategies for doing so are complex. Past attempts at eliminating child labour in, for instance, Bangladesh or Pakistan, by merely banning children from work, have created severe problems (Hussain-Khaliq 2004; Khanam 2006; Nasir Uddin et al. 2009; Siegmann 2008). In Bangladesh it was found that the result was to push children, particularly girls, into even more risky occupations, such as sex work (Nasir Uddin et al. 2009). A subsequent ILO project combined the elimination of child labour with access to education and income support for the family, which was significantly more successful. A similar approach combining regulation with income and education support is necessary to eliminate household-based child labour in India, but it should also be clear that a large part of the roots of the problem lie in the organization of the production networks and the functioning of value chains, which also result in very low incomes for home-based work. The challenge is thus not only to develop public policy interventions to deal with the roots of the problem in poverty and education, but also to develop public and private forms of regulation to govern local and global production networks. The further challenge resides in how to make these forms of regulation compatible, in a situation where incentives for the evasion of public regulation have thus far ensured the persistence of child labour.

Production network initiatives

Our analysis has shown that child labour in garment production continues largely at the household level, either in home-based production or household-based enterprises. We take the case of home-based production first. When work is taken into the home, there is an almost inevitable tendency for children to get involved in the work. Initially it may be just to meet deadlines in peak seasons, but over time the involvement increases and could result in school drop-outs. Home-based work has the advantage of providing flexibility to women. They can work for brief periods of time, say, after children have gone to school, then after lunch, and so on. This advantage of flexibility, however, goes along with the very serious disadvantage of promoting or providing a situation conducive to child labour. Further, it is difficult for any agency to monitor home-based work.

Is it possible to fashion workplaces that are near the residence? These would have the advantage of giving the women flexibility and eliminating travel costs. If the work place could be combined with a crèche, then the withdrawal of particularly girls to look after younger siblings, could also be eliminated or, at least, reduced. The Mewat project[8] tries to do this by providing a community space within the village or town where women performing embroidery and embellishment functions can come and work in a flexible manner. They usually work in groups for about two hours each before and after lunch. As a rule, they are not allowed to take work home, although in discussions some of the women did say that they occasionally did this. Child labour has nevertheless been kept out of the process by moving the work place to a community location that can easily be monitored.

The further key problem for the regulation of home-based production concerns various legislation issues. The first is that most forms of home work in India are excluded from coverage by national labour laws (see Deshingkar 2009: 10–11), and the second is that child labour is currently permitted in this setting under Indian law. There is a clear need for greater protection for home workers in general and closing the 'loopholes' in the Child Labour Act, which have in fact acted to favour its persistence. Again, the complexities have to be recognized: simply banning children from work in the home will be ineffective and counterproductive. The key, then, is legislation that takes as its basis the Right to Free and Compulsory Education Act and seeks to address child labour within that framework, but which clearly amends the exclusion of the home from definitions of what does and does not constitute child labour. As noted already, in 2012 the Government of India announced that it will amend the Child Labour Regulation Act to remove the exemption given to children employed within the family, although, at the time of writing, what will be covered under family and employment has not been clarified.

Household-based enterprises are quite different from home-based production, although it is clear that the aforementioned amendment to the Child Labour Act would go a long way towards complicating the possibilities for employers and recruiters to retain children as workers in these settings. Without the possibility of claiming that the children are relatives and that the unit is thus in compliance with the legislation, it seems likely that it would become much more difficult to conceal child labour. Yet this would clearly depend on effective systems of regulation in order to monitor the household sector. Labour inspection systems of some kind clearly need to be extended to cover informal units and workers, but, as Deshingkar (2009) points out, putting effective systems in place would depend entirely on much greater political commitment from both national governments and the private sector than has hitherto prevailed. Deshingkar also proposes that the ineffectiveness of conventional labour inspections calls for a different approach, where 'multi-party' inspections would involve the media, researchers, non-governmental organizations (NGOs), advocacy groups and human rights organizations. There are clearly questions of feasibility and organization involved in such a proposal, as well as questions of authority and legitimacy, but it is nevertheless valuable in highlighting the need to think very differently about public and private regulation of labour standards.

This returns us to the question of responsibility for compliance. The principal employer, as the notion exists in Indian contract labour law, is the final buyer of the service, that is, the global or Indian companies who pay for child labour. This contract labour law puts the onus of checking on child labour on the principal employer. This is also in line with notions of corporate responsibility, which is now generally accepted to extend down the value chain, a responsibility that cannot be side-stepped by recourse to contracting out work. The imperative is, therefore, towards the development of systems of regulation capable of dealing with the systems of outsourcing that characterize production networks, along with greater regulation of recruitment and labour contracting systems.

Along with the reorganization of women's home-based work in community centres, the elimination of the various subcontracting layers of intermediaries could increase income earned at the level of the household producer. Even without an increase in rates paid, the elimination or reduction of these numerous layers could increase women's income from embroidery and embellishment work. For instance, at Mewat, where the buyers directly give work to women, from which the organizing NGO takes 20 per cent, the women still earn Rs.15 per hour,[9] which even over a six-hour working day would give a daily income of Rs.90. This is just above the $2 per day poverty line income and is about 30 per cent higher than the per capita income in households with child labour in this sample. Of course, to lift a household out of poverty, adult earnings would have to be somewhat more in order to provide an above-poverty per capita consumption level for the household as a whole, including children who would now not earn an income.

In any case, eliminating or reducing the various layers of intermediaries would be the first step in increasing earnings from embroidery and embellishments and bringing them up to the prevailing minimum wage. The next step would be to bargain with the buyers in order to increase piece rates to cover not just minimum wages but provide a 'decent wage', which would largely cover family expenses.

Since we are dealing with global production, with the necessity of being competitive in the global economy, what would happen if wages for embroidery and embellishments increased? Full discussion of this issue would take us somewhat beyond the scope of this study. But it should be pointed out that the pressure of competition is frequently put forward as the reason for resisting any kind of improvement in working conditions, including the elimination of child labour. Keeping wages low, however, is not the only way to increase competitiveness.

That, in fact, is what is called the 'low road', on which other producing countries with even lower wages can easily overtake. Indian capital and trade unions, too, often portrayed the anti-child labour campaigns as a way to undercut the competitiveness of developing countries. There were certainly protectionist intentions on the part of some of the anti-child labour campaigners. But does that rule out the desirability of the action itself?

In brief, it can be pointed out that an increase in wages would increase pressures to increase productivity and also to move into higher value-added areas of production, as is in fact happening in China. These are not undesirable developments and, in any case, an increase in wages is a necessary condition for moving out of the completely unacceptable conditions in which garment home-workers and their children live. Eliminating child labour would increase the pressures for a movement from the low road of competitiveness based on subhuman living and working conditions, to the high road of competitiveness based on higher productivity and higher value-added products.

Social policy: employment guarantee or conditional cash transfers?

An increase in household income of below poverty line households could be brought about by government intervention. An urban counterpart to the National Rural Employment Guarantee Act has been under discussion for some time. Assured employment at prevailing minimum wages for 100 days would not only supplement household income, but would also raise the floor of urban wage rates. Of course, in such a move there would be budgetary implications. But in some combination of private sector and government initiatives, measures are clearly needed to increase household income at the bottom of the urban working class.

Yet we must caution against simplistic assumptions regarding the roots of child labour in poverty, inasmuch as our survey shows that there are households with child labour that are above the poverty line. This is in line with findings in research associated with the same project from which the present study emerged, this time in the agricultural sector in Brazil. There, a key finding was that it was not the poorest of the poor who were most vulnerable to forced labour, and consequently that government programmes to address extreme income poverty were likely to miss the target sections of the population, in turn further increasing their vulnerability to highly exploitative work (Phillips and Sakamoto 2012; Phillips 2013). Increasing household income, therefore, is only one part of an appropriate response and not applicable across the board as a diagnosis of the problem.

Could conditional cash transfers (CCTs) – cash transfers to women on the condition that children attend school and do not work – be more effective in reducing the incidence of child labour?[10] CCTs targeted at eliminating child labour have been tried in a number of Latin American countries, notably Mexico and Brazil. They have not been uniformly successful in reducing the incidence or working hours of child labour. Even so, some lessons have been drawn from these experiences (Gee 2010). The first is that the amount transferred should be sufficient to provide a real incentive to withdraw the child from labour and send the child to school. If the amount transferred does not make much of a difference to household income, then the CCT may not work. Second, the amount may need to vary with both gender and age – more to be paid for girls to go to school and for older children to continue in school. Third, as in most such cases, the money would need to be paid to the mother, whether or not she is the so-called head of the household. As many studies have shown, women tend to spend more of the money they control for their children's education and health, than do men (e.g. Murray 2008). Finally, as in Brazil, the household receiving the money should sign a contract with the state to send the children to school.

But will the children go to school and then work at home, neglecting their studies? Our study has shown that such combining of school with after-hours work is likely to lead to early drop outs as student's school performance suffers. So, it is important to eliminate such after-hours work. In Brazil this seems to have been substantially done by increasing the hours of school-based after school activity (Gee 2010). Forms of tutoring and recreational activities, organized at the community or school level, could help substantially reduce if not eliminate students' after hours-work. By contrast, in India there is a lack of opportunities for play and entertainment, particularly for girls, in the semi-slums in which they live and work. In the case of girls, this is as much a problem of the prevailing concern for safety, which makes parents limit their mobility, as it is the result of the more general lack of appropriate space for recreation and physical activity.

Eliminating child labour, however, is also a matter of changing social norms, which could be supported by the CCT. A study of Bihar (Datta and Rustagi 2010) showed that there is no social stigma attached to withdrawing a boy from school and sending him off to contribute to family income as a migrant in a distant city. It is even seen as a form of desirable transition to modernity, an escape from the village to the city. Thus, changes in the social norms of parents and the communities within which they live are also needed to eliminate child labour.

In a sense, economic capacity and norm changes are both required to eradicate child labour.

The point here is thus not to punish household-based violations, but to facilitate ways of complying with the requirements of sending children to school and eliminating child labour. This is the application of the general 'compliance' approach to regulation, where obligations cannot be discharged by paying a penalty and regulators work with all involved to find solutions (Piore and Schrank 2006).

Conclusion

Both global production and, even more so, production for the Indian domestic market extend beyond a number of firms and middlemen (contractors) to go right down to the household. Home-based production, whether in women's piece rate work or in household enterprises, in the tasks of embroidery and embellishment, is the lowest rung of the production network for the characteristic North Indian garment with embroidery and embellishment. In this home-based production, this study has shown that children are clearly and substantially involved in the production process. With below-poverty incomes and home-based production, children are drawn into working, first after school hours as helpers to meet deadlines in peak seasons and, then, dropping out of school to become full-time workers. Monitoring by global buyers and their agents rarely reaches down to the household level to see what is actually going on inside the house. Child labour consequently needs to be understood not only in terms of its roots in poverty, but also as the direct result of how contemporary production is organized and how value chains function in both local and global markets.

Addressing child labour consequently involves coordinated action across a range of areas and by an array of different actors. In the first instance it relates to the governance of production processes, both in terms of regulating social compliance and in terms of restructuring production networks such that home-based work is no longer entirely unregulated and unprotected, as it is at present. Enabling and empowering women to get higher incomes from their work, and moving such work out of the home into community centres near residences and under supervision, are further methods that could work to eliminate household-based child labour in garment production for global markets. A more general, state-based approach is that of a well-designed CCT combined with the already existing universal and compulsory education provision of the recent constitutional amendment. All this, however, needs to be combined with a law that prohibits child labour in any establishment, whether or not run by 'family'. A change in the Indian child labour law and a shift in the regulatory mechanism away from penalizing to finding solutions, along with CCTs and compulsory education, could form an effective combination for ending child labour in India.

Notes

1. This chapter arises from work conducted for the research project 'Vulnerable Workers in Global Production Networks: Case Studies of Trafficked and Forced Labour in Brazil and India', which was funded by a research grant from the Chronic Poverty Research Centre (CPRC) and led by Nicola Phillips. The study was carried out by the Institute for Human Development, New Delhi. The authors would like to thank Sam Hickey, Andrew Shepherd, Atul Sood, Stephanie Barrientos and Anne Posthuma for their helpful input at various stages of the project. We thank numerous colleagues from the Institute of Human Development, in particular Preet Rustagi, Amrita Datta, Balwant Mehta, Sheila Bhalla and Alakh Sharma for comments at various times. Sunil Kumar ably coordinated the field team, which also included Shailesh Kumar, Ramanand Jha and Madan Jha. We are also grateful to Stephen Buzdugan for additional research assistance. The present chapter draws on material published in Bhaskaran et al. (2010) and Phillips et al. (2011).
2. It should be noted that these figures reflect the National Sample Survey Organization's somewhat restricted definition of child labour as children working for a wage in the labour market. If we were to include those children who are neither at work according to this definition, nor at school, but in fact working with their families in some form or other, then the incidence is much higher. In 2007–2008, it stood at 15.4 per cent, having declined from 31.8 per cent in 1993–1994 (IAMR 2011).
3. A GPN is defined as 'the nexus of interconnected functions and operations through which goods and services are produced, distributed and consumed' (Henderson et al. 2002: 445).
4. Kanbur's four conceptual categories are (a) regulation applicable and compliant; (b) regulation applicable and non-compliant; (c) regulation non-applicable after adjustment of activity; and (d) regulation non-applicable to the activity (2009: 33).
5. The urban villages of Delhi have more than five or six times the population density of the non-village areas. They cater to mixed land use, with residential, commercial and industrial (small or household industries) within the same area, based on the demands of the surrounding localities.
6. ILO Convention No. 33 of 1932, Article 3, Para 1. Thanks to Benjamin Smith of ILO-IPEC for providing this reference.
7. In late 2012 the Government of India introduced legislation to amend the Child Labour Regulation Act in order to remove the exemption given to children employed within the family.

8. Mewat is a particularly dry and poor district of Haryana on the border with Rajasthan. The population largely belongs to the Meo community, who are also Muslims, though there are also some of the usual caste groups. In the late 1990s, an International Fund for Agricultural Development project in the area organized microfinance groups of women's Self-Help Groups. The NGO in the project has continued to work in the area and now organizes the embroidery and embellishment project. This pilot project is supported by the Government of India's Ministry of Rural Development, along with major international buyers Gap and Impulse (an Indian firm supplying to numerous international brands). The objective is now to develop this pilot project into a major one covering up to 20,000 women. With training in embroidery and embellishment work, and women working in community centres within their areas of residence, the hope is to develop Mewat as a brand for 'child-free' embroidery and embellishment. With the ubiquity of home-based work by women in embroidery and embellishment, it is interesting to note that, among all its suppliers worldwide, Gap makes an exception allowing home-based production only in North India. While enabling women to retain the advantage of flexible working hours in home-based production, it is certainly necessary to move to eliminate child labour in the process – something that could possibly be done by setting up community-based work and crèche centres within, or near, residential areas.
9. The piece rate is calculated on the basis of work that would provide a daily income of Rs.160, which is the minimum wage in Haryana, of which Mewat district is a part. Of this wage, 20 per cent goes to the NGO that organizes the community space for working and supervises and facilitates the process. The buyers have detailed calculations of the time required for each type of embroidery or embellishment work. There would of course be struggles over the correctness of these time calculations, but the principle is that piece rates should be in line with the daily minimum wage for an eight-hour working day.
10. It is also possible, as argued in Hanlon et al. (2010), that rather than expensively implemented CCTs, simple cash transfers may be quite adequate.

8

Workers' Perceptions of Compliance with Labour Standards: Assessing Opportunities and Challenges for Better Work in Lesotho's Apparel Sector

Kelly Pike and Shane Godfrey

Introduction

The apparel sector has been a springboard for industrial development in many countries. The sector, however, has also been notorious for poor working conditions and rampant exploitation of mainly female workers. National systems of enforcement have proved inadequate to the task of ensuring compliance with legislated standards and trade unions have generally made little headway in the sector in developing countries. The reputational damage this situation caused to major brands and retailers in the United States and the European Union (EU) has resulted in efforts to develop private governance to address non-compliance, mainly through the mechanism of codes of conduct and buyer audits under the rubric of corporate social responsibility (CSR). The Better Work programme grew out of an alternative approach first piloted in the Better Factories Cambodia programme. Since then, Better Work has been rolled out in an additional six countries and has emerged as a leader in the social regulation of international labour standards. Lesotho is one of these countries.

Lesotho is a small country with a population of less than two million, the majority of whom are engaged in subsistence agriculture (Silici 2010: 7). It is one of the poorest countries in Africa with very high levels of poverty, unemployment and HIV/AIDS. There is little industrial development: for years the main source of income was remittances from its male citizens employed as migrant labour in South African mines (Murray 1981: 86). The apparel sector is a relatively recent phenomenon and it is vitally important for job creation, poverty alleviation and economic development.

The apparel sector currently comprises 45 firms employing about 35,000 workers. Approximately 25 of the firms are owned by Taiwanese investors who export to US brands such as Gap, Levi's, Kmart, Childrens Place and Kohls. The remaining 20 are owned by South Africans who export back in to the major South African retailers. The former, which tend to be larger, are clustered in Maseru, and the latter in Maputsoe. There are no locally-owned firms. These two value chains had quite different starting points and motivations (documented in more detail by Morris et al. 2011).

Both sets of manufacturers face intense competition and are subject to downward pressure on labour standards and working conditions. The enforcement capacity of the Lesotho government is reportedly limited and inefficient; it is a poor second best to buyer audits. And, with the continued importance of Lesotho in global and regional value chains, there is a real need for both economic and social upgrading assistance in both value chains. It is in this context that Better Work Lesotho operates.

A considerable literature has emerged over the past 20 years documenting, and at times criticizing, the efforts of retailers and brands to ensure decent working conditions in global supply chains that extend into developing countries (Doorey 2008; Elliott and Freeman 2003; Esbenshade 2004; Frank 2008; Gereffi and Mayer 2006; Gereffi et al. 2005; Jenkins 2002; Locke et al. 2009; Locke et al. 2007; Mamic 2004; Mayer and Pickles 2010; Pearson and Seyfang 2002; Yimprasert and Candland 2000). More recently, several scholars have analysed the impact of Better Work on supplier compliance, mainly in Cambodia. The research includes that presented in Brown, Dehejia and Robertson in this volume (Chapter 10) as well as work on the relationship of compliance with competitiveness (Oka 2012), the impact of Better Work on wages (Robertson 2011), factory survival (Brown et al. 2011) and human resource management innovation (Robertson et al. 2011).

This study seeks to contribute to this literature but differs from it in a number of important ways. First, the research began only five months after the official launch of Better Work Lesotho, so it was too early for this to be a before and after study of the impact of the programme. Rather, it is a study of the work conditions at the outset of the programme, allowing for an assessment of the opportunites and challenges that Better Work faces in Lesotho. The chapter focuses on the unique set-up of the two value chains in Lesotho's apparel industry, and how work issues vary between the two. The authors also assess whether

Better Work's comprehensive assessment tool captures the issues that are relevant to workers in Lesotho, as reported by the workers themselves. In line with this, it is possible to assess whether Better Work's efforts at improvements are aligned with what workers think will make their work better.

Second, the study engages primarily with workers regarding their perceptions of labour standards, working conditions and compliance, as well as seeking their comments on what can make their work better. By placing the worker stakeholder at the centre of the discussion, the study provides a springboard for future research that places workers at the centre of a comparison of the impact of Better Work over time.

Third, the presence of two value chains in the Lesotho apparel sector, defined by different ownership nationality and end markets, provides an opportunity to compare the perceptions of workers regarding working conditions and compliance with labour standards in each. While Taiwanese owners have been attracted to Lesotho primarily by the benefits of AGOA and export to the United States, South African manufacturers are motivated mainly by the much lower labour costs compared to South Africa (Morris et al. 2011). The hypothes is was that this would influence their approaches to managing their factories, which would in turn be reflected in workers' perceptions of what makes for better work.

Despite the aforementioned differences, the aim of the study is congruent with the existing research on Better Work: to understand what role Better Work can play in improving labour standards and working conditions. A second question is explored in the context of Lesotho, namely whether owner and management nationality leads to variation in workers' perceptions of what makes better work.

Compliance with prescribed labour standards and working conditions

In the literature on global governance and corporate responsibility that has emerged over the past two decades, the term 'labour standards' is often used interchangeably with 'working conditions'. Better Work has distinguished between the two by creating a model of compliance clusters, and it uses this model as part of its monitoring and evaluation practices. The first cluster is 'core labour standards'. This includes the abolition of child labour, forced labour, protection from discrimination in respect of employment and the right to freedom of association and collective bargaining. The second cluster is 'working conditions'. This includes compensation, contracts and human resources, occupational health and safety, and working time.

Previous research, based on evidence from Better Factories Cambodia, has reinforced the validity of these categories and has demonstrated their appropriateness in assessing labour standards compliance (Brown et al. 2011: 23; Robertson et al. 2011: 12). In this study, the feedback of workers was solicited on (1) the issues most important to them, and (2) what they believe can make their work better. The components of both compliance clusters (i.e. core labour standards and working conditions) are used to measure workers' perceptions of what makes their work better. In seeking to understand the key research question, 'what do workers think will make their work better?', we consider the following issues:

- Is there variation in workers' perceptions of compliance between the two value chains?
- Do the Better Work compliance clusters broadly cover the range of issues that matter to workers?
- What do workers want?

Methods for data collection and analysis

A total of 17 FGDs were conducted with 133 workers, with about eight to ten people per group. All the focus groups were digitally recorded. The focus groups were conducted in three waves (May 2011, September 2011 and December 2011) and participants were recruited by their unions. The participants represented 36 factories in both industrial sites, 14 of which had subscribed to Better Work. The gender ratio was three to one women to men. Their feedback was coded and organized according to the compliance clusters and was sorted using qualitative software. The software also assisted with running queries that helped to identify patterns in the data and allowed an integrated analysis to be conducted, treating the data both qualitatively and quantitatively (Bazeley 2002: 229). In addition to organizing workers' feedback by themes defined by the compliance clusters, new themes that emerged in the discussions were also coded and analysed. This created room for workers to express their feelings about aspects of working life that may not have been anticipated by the researcher but which play a role in shaping their perceptions of what makes better work. Workers were also asked directly to share their opinions on what could make their work better.

Guidelines for the FGDs were established following the model of the Better Work compliance clusters. This is a model that has been applied in research on other Better Work country programmes during the monitoring and evaluation assessment process. Research that draws on data derived from using this model is also emerging in the literature (Brown et al. 2011: 25; Oka 2012: 13; Robertson 2011: 29; Robertson et al. 2011: 15; Seo 2011: 31). Basing the guidelines on these established measures makes the results comparable, which will assist in future research comparing the Better Work programme across countries.

Table 8.1 Demographics of focus group participants by value chain

	Better Work		Non-Better Work		Total	
	Factories	Workers	Factories	Workers	Factories	Workers
T-USA	12	43	8	28	20	71
SA-SA	1	12	12	43	13	55
Unassigned	1	2	2	5	3	7
Total	14	57	22	76	36	133

Initially, nodes (or folders) were created for the five compliance categories already mentioned: core labour standards, compensation, contracts and human resources, occupational health and safety, and working time. New nodes were created for themes that emerged while proceeding through the transcripts. For example, the issue of workers' relations with supervisors came up often, so a new node 'supervisor relations' was created. Also, by establishing attributes for each participant in the beginning, the findings could then be sorted by participant after the coding was complete. This made it possible to analyse the data according to specific attributes (e.g. owner nationality, end market, Better Work (BW) status).

The Better Work Lesotho programme was in an early stage of implementation at the time this research took place (it was officially launched five months prior). Taking this into consideration, in addition to being questioned about their working conditions, workers were asked what they believed could make their work better. The following is a discussion of the findings.

Findings

The findings of the study are presented under the broad headings of (1) the Better Work compliance clusters (broken down by core labour standards and working conditions), and (2) Additional issues. Tables 8.2 and 8.3 provide a summary of the frequency with which different issues within these categories were raised.

Table 8.2 Ranking of issues within each value chain

	T-USA	SA-SA
Core Labour Standards		
Child labour	0	0
Discrimination	5	6
FOA and CB	3	8
Forced labour	1	3
Working Conditions		
Compensation	10	4
Contracts and HR	13	4
OSH	34	15
Working time	5	4
Additional Issues		
'Chinese' management	4	0
Government	3	1
South African management	0	9
Supervisor relations	18	34
Unions and collective power	1	11
Work life	3	1

Analysing workers' feedback according to whether they belonged to a Better Work-subscribed factory was not useful at the time of research, given that the programme had only officially launched five months prior to the first wave of FGDs. Though it was too early to draw conclusions about the impact of the programme at that time, workers' feedback has been distilled by Better Work status in addition to their global value chain placement, in order to serve as a point of comparison in future research on Better Work Lesotho. One can see from Table 8.3 that there is not a great deal of variation between Better Work and non-Better Work factories within each value chain. Interestingly, in several cases where there are discrepancies between Better Work and non-Better Work factories, the Better Work factories appear to fare worse.

'Supervisor relations' and 'OSH' (occupational safety and health) weighed heavily in the focus group discussions. Occupational safety and health was the number one issue in Taiwan-USA (T-USA) factories, and supervisor relations was the top issue in South Africa-South Africa (SA-SA) factories. These are discussed in more detail in the next section. Though issues related to core labour standards came up relatively less frequently, they shed light on different instantiations of non-compliance, particularly with regard to discrimination in the workplace. Where relevant we identify differences related to ownership nationality.

Table 8.3 Ranking of issues within each value chain, organized by Better Work (BW) status (%)

	BW T-USA	Non-BW T-USA	BW SA-SA	Non-BW SA-SA
Core Labour Standards				
Child labour	0	0	0	0
Discrimination	5	4	8	5
FOA & CB	2	2	10	11
Forced labour	1	0	0	5
Working Conditions				
Compensation	10	9	8	5
Contracts and HR	11	19	3	3
OSH	37	38	10	9
Working time	4	6	10	3
Additional Issues				
'Chinese' management	4	6	0	0
Government	2	6	0	0
South African management	0	0	13	7
Supervisor relations	20	9	31	41
Unions and collective power	1	0	5	10
Work life	4	0	3	2

Core labour standards

On core labour standards, the major concerns of workers related to freedom of association and also to discrimination – specifically with respect to pregnancy and HIV/AIDS.

Discrimination

There was consensus among the participants that men and women are treated equally in terms of the wages they are paid. Some men said they are discriminated against, stating that it is difficult for them to get jobs because factory managers 'don't want male workers' or prefer female workers because they supposedly 'know the machines and are easy to work with'.

Workers in SA-SA factories mentioned issues relating to discrimination slightly more frequently than workers in T-USA factories, but only workers in T-USA factories mentioned the issue of discrimination against pregnant women. Their feeling was that pregnant women are pushed so hard that they leave, rather than being fired outright. Gender discrimination was also discussed in the context of sexual harassment. For example, if a male supervisor is interested in a female worker, makes an advance on the woman, and she refuses him, he can treat her unfairly by giving her undesirable tasks or making work more difficult for her in some way.

Workers in both types of factories spoke about discrimination against people with HIV/AIDS. The Lesotho Labour Code provides that an employer shall not compel an employee or a job applicant to disclose his or her HIV/AIDS status or that of any other person (s 245 C.(1)). Nor may an employer directly or indirectly force an employee or job applicant to be tested for HIV (s 235 D. (1)). However, several workers indicated that it is still possible for the supervisors to determine one's status by requiring them to visit the factory clinic. People with HIV said it is easy for them to be identified because they require permission in order to get their medication or go for check-ups. Once people know their status, it makes them vulnerable to mistreatment.

> When one worker asks to go to a doctor, they want to see the book, they want to see exactly do you have a problem. If the supervisor learns from the doctor that there's an issue with you, if they know you suffer from this and this, they'll talk about it all the time.
>
> (Factory worker)

Freedom of association and collective bargaining

A compliance assessment in a factory might unearth whether or not there are unions in the factory, whether workers are allowed to join unions and what mechanisms there are for collective bargaining. An important underlying issue related to this topic, however, is the access that workers have to such mechanisms, including the real 'freedom' they have to join a union. Anti-union discrimination and obstacles to freedom of association were raised as an issue in eight out of the 17 FGDs. For example, some factories have automatic dues deduction from payroll while others require that workers come individually to the main office in order to pay their dues. For some, this is not a problem. For others, it is intimidating to do this in front of their bosses, and many let their membership in the union lapse for fear of negative consequences on the job. When automatic deduction is stopped, there is also a tendency for some workers to not pay their dues themselves.

Some workers reported that they are taunted or given unpleasant jobs as a form of discrete punishment. They described it as management pushing them to resign and showing them that the union cannot help them. In addition, once a worker is promoted to supervisor, they are discouraged either from remaining with the union or ever considering joining the union, and instead it is assumed that they are now 'on management's side'. Some felt that managers use this is a strategy to keep the power of the union at bay:

> When they see a person that he is strong at the union, they promote them to be supervisors, so he can be on their side.
>
> (Factory worker)

Problems regarding freedom of association and collective bargaining were raised more than twice as often by workers employed at SA-SA factories than T-USA factories. This possibly reflects a reaction by South African managers to their experiences in the heavily unionized apparel sector in South Africa and their desire to avoid the same thing happening in Lesotho.

Forced labour

From a legal point of view, not being able to leave early on payday is not forced labour, and mandatory overtime is not usually considered forced labour. In Lesotho, the legal limit is 11 overtime hours per week. All participating workers reported that their overtime hours were within this limit. However, when workers spoke about managerial practices, such as not allowing workers to leave early on payday or requiring them to come in for overtime when they had other obligations, they interpreted this as a form of forced labour.

> She [the supervisor] makes us stay for overtime even if we don't want to come, and on Saturdays when I want to go to funerals I am not allowed. I am supposed to come to work. They make it a must that we come to work for overtime.
>
> (Female factory worker)

There were not many of these complaints, though where they did exist they were mainly by workers in SA-SA firms.

Working conditions

Compensation

Presented with the topic of 'making ends meet', workers gravitated towards the key concern that their expenditures grossly exceed their income. They also frequently brought up two additional issues that are indirectly related to compensation. (1) There is inflexibility regarding doctors and sick leave, and (2) pay increases are sporadic and based on favouritism. Most of the compensation-related issues raised in the discussions (and in particular the first two key points) were seen by workers as linking directly to their health and safety. With the salaries they earned, they could not afford to properly feed their children and they were forced to take loans at extremely high interest rates (40 per cent). Discussions around the issue of compensation shed light on the implications it has on a range of issues both within and outside of the workplace.

Additionally, workers stated that they are not provided with compensation in the case of an injury on the job, and many workers found it difficult to take the 12 days of sick leave they are entitled to. First of all, it is difficult to provide the required proof (limited to specific doctors), especially if they usually visit traditional doctors. Secondly, they believe there is an issue of favouritism (e.g. a supervisor who 'does not like you' can keep you from returning to work if you have taken sick leave). The issue of sick leave falls within the 'working time' category in the Better Work clusters rather than 'compensation'. However, because workers see sick leave as a form of social security and/or benefit, it seemed to correspond much more closely with: 'compensation: social security and other benefits'. When workers spoke about working time it was with reference to their starting and ending times. 'Leave' in the 'working time' category, therefore, did not correspond to the sense of health and well-being that workers were attaching to not receiving adequate sick leave.

Workers also felt that individual pay increases are given according to an unwritten policy of favouritism. Many mentioned that there are workers doing the same job who have been there the same amount of time, yet one will receive a raise and the other will not. It is the same with annual bonuses. There will be years that go by without some workers getting a bonus, whereas others in an equal position receive theirs. It is not clear to employees what they can do to ensure they receive a raise or bonus, other than to try to get their supervisor to 'like' them. One worker exaplained: 'The supervisor is the one who goes to the manager and says you should give a raise to so-and-so. It's usually because they like them.'

Contracts and human resources

In the FGDs, workers' statements related to the topic of contracts and human resources (HR) could be grouped under three categories: HR relations, contracts, and the Directorate for Dispute Prevention and Resolution. HR relations received the most attention.

Across all factories, the HR officer positions are held by local Basotho. This was intentionally built into employment policy, in part to compensate for the language difficulties of non-English-speaking foreign managers. Aside from payroll and other logistics, HR is primarily responsible for dealing with disputes and discipline. When workers have an issue, they must first speak with their line supervisor and then their manager or production manager, but ultimately they can go to HR if the issue remains unresolved. This is a problem for workers who find it difficult to approach their supervisors. The problem is further perpetuated by a sense among workers that HR tends to side with supervisors. 'If there is a problem between a worker and a supervisor, and a worker goes to personnel, the personnel will take the supervisor's side', one factory worker said.

Workers also expressed dissatisfaction with the contracts they are given: they are often in English and not explained. Those workers who did receive contracts indicated that none exceeded one year in length but that they were renewable annually, making it easy for employers to let go of employees.

Workers are concerned that HR managers are promoted based on their ability to 'motivate' (or be harsh with) workers, and that they tend to side with management, brushing workers' issues aside. As one employee commented: 'When the other personnel got dismissed, the score lady was actually turned into the personnel. She doesn't talk good. She often harasses us. ... She's doing that work because of her dirty language.'

Complaints regarding contracts and HR were much more prevalent with regard to T-USA factories. This is possibly because SA-SA managers come from a heavily regulated labour relations environment and usually have considerable HR and industrial relations management experience. It is probable that this experience leads to closer monitoring of the local HR managers than is the case in T-USA firms.

Occupational safety and health

Many workers complained of strong chemicals being used in their work, which affected both the person using the chemical and workers in neighbouring sections. In some cases, unlabelled chemicals led to hazardous encounters for employees. One worker reported that a man had mistaken a chemical for water, and it burnt his insides. Another said that there is in-fighting between some of the workers, and she is afraid that someone might one day use that chemical water to poison her food. In addition to hazardous chemicals, several workers mentioned that the machines they use are unsafe and that they felt their bosses wanted to save at the expense of the workers' health. Some said they were using older (non-automatic) sewing machines because they are cheaper, which leads to more finger injuries. Others said they use equipment that has been modified in some way to make production more efficient.

> You would have instructions on the containers that this thing should be used appropriately like this, but then the Chinese, or our supervisors, will just instruct us to use it wrongly, simply for the convenience of achieving results.
>
> (Factory worker)

Some workers were given personal protective equipment, though most reported that it was either poor quality (i.e. the fumes or dust could pass through it) or provided only when buyers were visiting. Several people mentioned the issue of buyer visits. Workers who had at some point experienced a buyer visit, or received personal protective equipment, were more likely to express discontent with the quality of occupational safety and health protection. A possible explanation for this is the heightened awareness of what is possible when management is paying more attention to their working conditions. They were tired of the quick-fixes and most of their statements included the phrase 'only when the buyers come ...' or 'we know the buyers are coming when ...'. This concerned them because, 'once we know the buyers are not there, it's definitely a problem'.

This was an issue particularly among workers from T-USA firms. One possible explanation for this is that they are supplying the US market, where buyers are more likely to have codes of conduct that they come into factories to monitor. When workers see these buyers, and receive some kinds of improvements due to the buyers' visit, they are more likely to raise the issue of inconsistent if not poor health and safety standards.

Working time

Feedback from workers indicated that firms comply with the law on regular hours of work, which is 45 hours per week. Many workers expressed their concern that the 7 a.m. start time is too early. This is especially so in the winter when it can get extremely cold and it is still dark when they are walking to work. Workers stated that it is difficult to work under such conditions and that they would be much more effective if the working day could be changed to an 8 a.m.–4 p.m. shift, rather than 7 a.m.–5 p.m.

Another issue related to working time that frequently came up in discussions was that of workers' ability to meet their targets. Most felt that their targets were too high and that they struggled to reach them. Often, there were negative consequences for this. In some situations, workers opted not to take tea breaks or go to the bathroom, in order to maximize their working time.

Additional issues

In addition to using the compliance clusters framework to organize workers' feedback, several other issues were raised by workers throughout the course of the focus groups. These issues were outlined in Tables 8.2 and 8.3.

The issue that came up most frequently was that of 'supervisor relations', accounting for 18 per cent of all comments made by workers in T-USA factories and 34 per cent of all comments made by workers in SA-SA factories. Collective power is not a reference to freedom of association and collective bargaining, but rather to workers' desire for (1) unity between the unions, and (2) stronger collective power through this unity. Workers also expressed dissatisfaction with government's passive role in ensuring their wellbeing, as well as with managements' lack of concern for their work–life balance. This section focuses on the dominant issue of supervisor relations, which, it is argued, underpins most of the issues discussed in the focus groups.

Supervisor relations

The issue of supervisor relations came up in virtually all of the focus groups. In some cases, it dominated much of the discussions. Many of the things workers said about issues within the compliance clusters can be related back to the issue of supervisor relations. In fact, supervisor relations were the source of many of the specific issues that were raised. Some workers reported that male supervisors abuse their power by coercing female workers to go out with them. Others said that supervisors find ways to determine whether workers are HIV positive, and will discriminate against them on this basis. Many workers said that supervisors use favouritism, which affects who gets a pay raise, who gets sick leave and who gets selected to supervisory roles. When unskilled workers are promoted on the basis of being 'liked' by a supervisor, workers said this ends up creating more work for them, putting them under more pressure to meet their targets. Additionally, supervisors serve as the middle man between workers and HR, which left many workers feeling like there were obstacles blocking them from addressing their concerns.

Further investigation of the focus group feedback led to the discovery of additional sub-themes within the topic of supervisor relations. Specifically, these were broken down in to seven categories: conflict and power, disrespect, doctors and sick leave, favouritism, forced work, sexual harassment and skills/knowledge sharing. Table 8.4 illustrates the number of references made by workers that could be categorized within one of these sub-themes.

Table 8.4 Ranking issues related to 'supervisor relations' by value chain (%)

	Workers' ranking of issues in T-USA factories	**Workers' ranking of issues in SA-SA factories**
Conflict and power	26	38
Favouritism	26	18
Disrespect	22	18
Skills and knowledge sharing	20	14
Doctors and sick leave	2	6
Forced work	4	2
Sexual harassment	0	4

The sub-theme 'conflict and power' was established to capture what workers described as 'everything changing' when other workers were promoted to being a supervisor. The statements made in this subcategory reflect the divide between these two classes of workers, and the ways in which supervisors exert their control. Workers described supervisors as turning against them, not listening to their concerns and using petty issues against them as a form of control. Some suggested this is due to the pressure they are under to impress their managers, in order to hold on to their supervisory positions.

> Before then they are just on the same level with everyone else. Like you can eat together but once they're promoted you find that the person eats alone.
>
> (Factory worker)

The sub-theme 'disrespect' was created to capture workers' perceptions of feeling mistreated and talked down to in the workplace. Many reported that they are shouted at; others said they are treated like slaves. Several brought up the issue of young supervisors disrespecting their elders, speaking to them however they wanted to in the workplace. Though some of the statements were with reference to Taiwanese and Chinese supervisors, workers often spoke of issues with their local supervisors, with one suggesting that: 'I have a problem with the Basotho supervisors. They don't treat us well, and they treat us like lunatics when we talk to them.'

Another key issue for workers was that of 'favouritism'. The word 'favouritism' was among the top 25 words used most often in all references to supervisor relations. Favouritism was discussed largely in the context of promotions, pay increases and punishments or mistreatment. Workers felt that certain people received certain advantages from supervisors because they were either their family member, romantic partner or someone they 'liked' better than the others. Additionally, these informal selection procedures led to issues with skills and knowledge sharing.

> Workers who are ultimately elected, you find that basis is not that they have experience or that they are bright or something like that, it's just because you can talk whatever you like. So you get these people who don't know work, yet they supervise us.
>
> (Factory worker)

Workers felt they could not learn from their mistakes or that they spent too much time trying to show other workers how to do their jobs. In both cases, this was a result of supervisors not having the knowledge to show them how to do their work better. This slowed down production on the line, which led to managers putting increased pressure on the line supervisors. Inevitably, it was the workers who ended up feeling the brunt of this pressure.

What our findings suggest is that local supervisors mediated between foreign owners and workers in a way that significantly watered down any differences in management style, instead inserting a much stronger and more problematic relationship with workers.

Foreign management

When the research began it was unknown what to expect with regard to workers' perceptions of the two different sets of foreign owners/

managers. Preliminary research suggested that workers in T-USA factories would have very negative perceptions of management and working conditions. Hart's research on Taiwanese apparel manufacturers in South Africa refers to them as complaining continually about their African workers' 'low productivity', which was attributed to 'laziness, low levels of education, and ingratitude'. Not surprisingly, Hart found that workers were deeply resentful of their treatment by Taiwanese managers (Hart 2002: 166–67, 190–91). Furthermore, research by Lee points to stark differences between mainland Chinese managers and African workers with regard to attitudes to work that emphasized difference and stoked antagonism (Lee 2008). Although Hart rejects cultural explanations for the nature of the relationship between Taiwanese managers and African workers, it is difficult not to acknowledge that culture plays an important part (Hart 2002: 168–69). But it is not just about culture or power. Lesotho is still a predominantly agricultural country. In many cases, the Basotho workers entering an apparel factory in Maseru or Maputsoe are entering a factory for the first time. The factory environment and the intense discipline that goes with it is an entirely new and alien experience. It was, therefore, anticipated that, for whatever the cause, workers would generally view their Taiwanese managers in a poor light. In one of the focus groups, a worker reinforced this expectation when she said:

> The Chinese do not work with people nicely. They are rude. It's like they have been trained to talk to Basotho in a certain way, not that they don't speak English or Sesotho. They just speak to people as they please.
>
> (Female factory worker)

On the other hand, the relations under apartheid between white South African managers and black South African workers have been well documented. A broad description of these relations and management style is probably best captured by Von Holdt's concept of the 'apartheid workplace regime', which:

> consisted of a racially oppressive order derived from South Africa's settler-colonial history. Like other spheres of society, the workplace was the site of racial domination buttressed by racial segregation, and by racist discourses and practices in which the distribution of occupations, skills, incomes and power was racially defined.
>
> (Von Holdt et al. 2005: 7)

However, the strength of South African trade unions and progressive labour legislation have forced South African managers to radically change their management style. Many are now fairly comfortable dealing with unions, have become used to operating within the constraints of recognition agreements and labour legislation and are accustomed to negotiations with unions. This perspective was confirmed by one worker's comment: 'Our [South African] manager has allowed us to have a committee who will represent us and talk about all issues.'

However, the SA-SA apparel firms have located to Lesotho primarily to get the benefit of lower labour costs and a much more weakly organized sector than they are used to in South Africa. It is arguable that in this situation some South African managers would slip into old habits and think they had a free hand to drive workers as hard as possible. Although there is not a great deal of research on this issue, Miller's study of the Shoprite retail store in Zambia indicates that when South African management enter the region they take with them a strong sense of superiority, which has a racial dimension and is also rooted in a belief that they are technically more advanced and experienced than managers in less developed countries (Miller 2005).

A mix of responses was found. Some workers alluded to the attentiveness of their South African managers who allowed workers to come and talk to them about their issues. On the other hand, it was found that a number of workers in SA-SA factories had an intense fear of their South African owners. It was not entirely clear why this was the case. Many of the South African owners are not at the factory every day. Some live in Durban and come to the factory once every few weeks, while others live just across the border in South Africa but leave the daily operation of the facility to their mid-level managers. Perhaps it is this remoteness that induces workers to view the owners as omniscient; workers are frequently warned when the 'boss' is coming to the plant and told to keep their heads down. Some workers said that they fear their South African owner so much that they hide when they see him coming. They believed that the owner can fire them on the spot, instantly, without any notice.

The different nationalities of the owners and managers, however, did not have a noticeable impact on working conditions. We found that workers in both types of factories were dissatisfied with their working conditions. Importantly, although some of the issues varied slightly between the two types of factories, the overarching theme in the T-USA and SA-SA factories was that it was relations with local supervisors, rather than foreign management, that underlay much of the dissatisfaction and could be seen behind complaints regarding other issues.

In the T-USA factories, there is generally one Basotho supervisor appointed per line, and one Taiwanese or Chinese supervisor for every two or three lines. The Taiwanese or Chinese supervisor is in charge of the technical aspects of work (e.g. setting up a new style, showing workers a complex task) and the Basotho supervisor is in charge of 'motivating' the workers. This arrangement, whereby Basotho supervisors are selected to push staff, sets them up to be especially harsh in the way they exert control over workers. This environment makes workers fearful of admitting to a mistake or asking a question, because they are worried they will be penalized.

In the SA-SA factories there is generally one supervisor per line, and they are almost always Basotho. There are a similar set of complaints about supervisors in these factories. Workers described promotion procedures as being random or based on favouritism and said that Basotho supervisors regularly abused their positions of authority. In addition, workers felt that these supervisors often had little experience, which made it difficult for them to properly learn from their mistakes.

What seemed to lie at the root of this situation was the relationship between foreign management, both Taiwanese and South African, and Basotho supervisors. Both sets of managers tend to have an arms-length approach to the Basotho supervisors and the factory floor. There is limited interaction between managers and supervisors before the supervisors take to the floor and begin overseeing the workers; and managers spend relatively little time on the factory floor monitoring the interaction between supervisors and workers. Whether this is because of cultural or language reasons, supervisors are to a large extent left to their own devices in managing the lines. This situation, combined with the lack of training given to supervisors, provides a fertile ground for them to abuse their positions. The range of problems that this gives rise to – rudeness, favouritism and so on – clearly has a negative impact on workers and productivity. Workers are reluctant to approach supervisors for assistance, so they turn to each other for help; many shop stewards reported that they spend a lot of their time showing workers how to do things or fixing their mistakes. This inevitably leads to production slowing down.

As one would expect, this has a negative impact on workers' morale. They are punished easily, and in some cases squeezed out of their jobs (e.g. when a supervisor wants to hire someone they prefer). As noted already, some workers said that the increased pressure to meet high targets has resulted in their not taking tea breaks or going to the bathroom. This can have serious consequences for their health, as well as detrimental consequences for performance.

Unions and collective power

In the focus groups many workers made reference to the unions in Lesotho's apparel sector not working well together. Some spoke about the competition for power that goes on between the five unions, fighting for their own membership rather than the collective benefit of all workers. Others mentioned in-fighting within one of the unions, whereby mixed messages are given to workers, leaving them confused about the union's overall agenda and/or their involvement in specific instances of collective action.

> Workers get confused because he's [Union X] but [Union Y] union leader will talk to the workers and say nothing is going to happen, so continue with work ... Some will go with the other union leaders, others will not do anything, they'll listen to the person who says nothing is going to happen. So it divides them and they get confused.
> (Factory worker)

Workers also expressed criticism about their fellow workers, arguing that they are too passive and too protective of their own individual security, rather than fighting for the rights of workers as a collective group. What they wanted was for the unions to come together so that workers could be on the same page and have a stronger voice. Workers in Maputsoe (where the SA-SA firms are clustered) were especially tired of the in-fighting and the mixed messages. Moreover, they were growing increasingly frustrated as they saw the workers in Maseru (where the T-USA firms are clustered) joining together under their unions. They felt restricted by the perceived attitude among the workers in Maputsoe that 'half a loaf is better than no loaf at all'. Under these conditions, organizing workers was difficult, as one employee explained: 'Even if we try and unite and say let's form a union so that we can be able to approach our managers, it will only be 10 people out of 100 people.'

Conclusion

Better Work assessments cover a broad range of workplace issues and, indeed, when workers spoke about the issues most important to them, several fell within the categories against which compliance is assessed by Better Work. Within the discussions around these issues, however, an important theme emerged. Namely that the issue of supervisor relations underpins much of what workers are dissatisfied with in the workplace. In talking about core labour standards, it was often the supervisors who were discriminating against workers on the basis of pregnancy or their health status, and the supervisors who workers felt were forcing them to stay for overtime. In talking about working conditions, workers recounted stories of unskilled supervisors getting in the way of their own learning on the job, preventing them from speaking to HR about their issues, berating them or not allowing them to take breaks for their medication. Most of what workers said, whether it was about provisions for their health and safety, their ability to meet their targets or their overall morale at work, could be traced back to the issue of supervisor relations.

In some cases possible reasons can be suggested as to why there were more complaints about a certain issue by workers at SA-SA firms than by workers at T-USA firms (e.g. freedom of association) or vice versa. However, there was not a consistent pattern of non-compliance by one group of factories as opposed to another. This chapter would argue, however, that supervisors are critical mediators between foreign owners and workers and that workers' views on many of the issues are much more strongly influenced by their relations with supervisors than the management styles of owners.

Better Work has taken steps to train supervisors in Lesotho's apparel sector, based on some feedback about supervisors needing training in people skills. It is important to note, however, that the needs run much deeper than how to interact with workers. The lack of a formal selection procedure is leading to instances of favouritism and bribery, lending itself to a further divide between workers and supervisors, and has implications for the relationships between workers as well. Their lack of technical skills slows down production, as workers cannot turn to them for advice. Additionally, workers would not want to turn to supervisors because they find them generally harsh to deal with and work in fear of punishment.

Workers felt strongly that training of supervisors would significantly improve their working conditions. Additionally, they believed there should be training for top managers, shop stewards and workers as well. Increasing awareness of one's rights, as well as educating their superiors on how to interact with them, was a combination of measures that employees felt would make their work better.

The level of detail collected in focus groups allowed for a nuanced understanding of the root causes of the most important issues for workers. By including workers in their full capacity as worker stakeholders, researchers and policy makers in CSR and international labour standards will be better informed and opportunities for improvements will increase. Additionally, workers are able to exercise their dignity by giving direct input on what can make their work better. As one worker eloquently ended a focus group discussion:

> It is helpful to discuss these issues because it raises awareness on the fact that we should know our rights, and we should know why we wake up in the morning and go to work, and also so that we could reach our goals at the end of the day.
>
> (Factory worker)

9

Workers' Agency and Power Relations in Cambodia's Garment Industry[1]

Dennis Arnold

Introduction

In the past three decades, a major transformation has occurred in the global economy as supply chains originating in 'core economies' have further expanded their sourcing networks into new frontier regions of production across East and South-East Asia (Arnold and Pickles 2011; Gereffi 2005). Scholarship on this phenomenon has expanded accordingly. Global value chain analysis describes the functionally integrated but geographically dispersed range of activities that firms and workers do to bring a product from its conception to its end use and beyond (Gereffi 2005). GPN analysts have critiqued the global value chains framework and its chain metaphor that assumes an invariably vertical and linear sequencing (Coe et al. 2008). In their terms, GPNs are a tool to conceptualize intricate linkages formed through multi-dimensional layers of economic activity. Thus, GPNs are better able to understand firm-centred production networks and the concrete political economic contexts in which they are embedded. However, absent from both frameworks is labour as agency (Cumbers et al. 2008). Selwyn (2012) argues that the most significant weakness of the global value chain approach is firm-centrism and the failure to conceptualize capital–labour relations as co-determinates of economic development. Drawing on Coe and Jordhus-Lier (2010), this chapter contributes to GPN analysis by demonstrating that the potential for worker action should always be seen in relation to the formations of capital, the state, the community and the labour market in which workers are variably embedded.

The chapter focuses on whether garment factory workers in Cambodia have the collective voice and ability to negotiate a living wage. Workers' agency is examined through a case study of a large-scale strike in

September 2010 over national minimum wage negotiations, led by two Cambodia trade union federations. Analysis is centred on four structural impediments to workers' wage demands. First, the Cambodian People's Party consolidated power in 2008 and is alleged to have rigged National Assembly elections. As a result, space for independent trade unions and civil society is constrained. Second, Cambodia is not deemed 'competitive' as a global sourcing option in terms of price, quality and speed to market. As a result, low wage and proliferation of unmonitored subcontract factories are increasingly the industry's competitive advantage vis-à-vis Bangladesh and Viet Nam. Third, the proliferation of fixed duration contracts in Cambodia means work is less secure, with attendant impacts on workers and unions' negotiating strength. And fourth, the unusually high number of plant level and national trade union federations makes it difficult for 'genuine' unions to promote the rights of their members, and workers' agency potential is marginalized. The intersection of these four structural forces circumscribes workers' and independent trade unions' ability to rework power relations with the employers association, the Garment Manufacturers Association of Cambodia. Despite the challenges, workers and independent unions have come to recognize themselves as the agents who must shape key demands including wage.

The chapter is organized as follows: the section below provides an overview of the garment industry in Cambodia, its significance and its governance structure. The following four sections address, in turn, the formations of the state, capital, labour market and community in which Cambodian workers are embedded. These four sections are followed by a case study on the 2010 garment workers' strike in Phnom Penh and its vicinity. The central contention is that Cambodia's garment industry is a paradox – it is characterized by a highly confrontational industrial relations environment while maintaining an extensive workplace monitoring programme. It is promoted as a 'showcase' for ethical production, yet tripartite dialogue is dysfunctional with few institutional means for workers to engage stakeholders, including trade unions, on the concern that is most pressing for any garment factory worker seeking to alter distributional relations in capitalism – wage.

Garments in Cambodia: an overview

Cambodia's four economic 'growth pillars' are textiles and garments, tourism, construction and agriculture. The garment industry directly employed 350,000 workers in 2008. Employment dropped to 296,800 in

2009 during the economic crisis, rising close to pre-crisis levels by mid-2011. In the first quarter of 2012 there were 356,609 workers in 320 registered textile and garment factories (BFC 2012). The Economic Institute of Cambodia (2007) estimates that direct production-related employment accounts for only 53 per cent of employment generated, meaning total employment related to this sector may exceed 600,000, significant for a country of some 14 million people.[2] Some 85–90 per cent of textile and garment workers are rural–urban migrant women who contribute a significant proportion of their earnings to their families in rural areas, who comprise roughly 80 per cent of the country's population (Economic Institute of Cambodia 2007). This is increasingly critical, as landlessness and land poverty have reached epidemic proportions. Of roughly 3,900,000 hectares of arable land in Cambodia, Licadho (2012) research shows that the government has leased 2,033,664 hectares of land to private companies under its land concession schemes – approximately 800,000 hectares in 2011 alone.[3] This increases workers' and their families' dependence on urban and peri-urban labour markets.

The first factories producing textiles and garments for export opened in Cambodia around 1994, with investors from Hong Kong (China), Malaysia, Singapore and Taiwan (China). The industry remains dominated by foreign investors, with Cambodian investors accounting for only 7 per cent of ownership in 2008. By comparison, the majority of firms in Bangladesh and to a lesser extent in Viet Nam are locally owned, which increases economic upgrading possibilities and the potential for local linkages and spill-over due to more decision-making power located locally (Staritz 2011). The Garment Manufacturers Association of Cambodia (GMAC) had 284 members in July 2011, down from its peak of 330 in 2007. Ken Loo, general secretary of GMAC explained that this is due primarily to consolidation (interview 11 July 2011). Membership is required of all exporters. The fieldwork shows that top buyers currently include retailers and big box stores, primarily US and European companies, in addition to East Asian intermediaries. Gap and H&M have long been the two most important, with other prominent buyers including Levi's, Li&Fung, Walmart, Inditex, Adidas, Target, Sears and VF.

Work conditions in the early years of Cambodia's garment production (1994–2000) were poor. Cambodia's labour relations were defined in the mid-1990s to mid-2000s by the state's structural inability or indisposition to implement labour standards (Arnold and Toh 2010). In June 1998, labour and textile manufacturer groups in the United States petitioned their government to review the alleged abuse of workers' rights in Cambodian factories (Polaski 2009). In 1999 the US and Cambodian governments initiated the three-year US-Cambodia Textile and Apparel Trade Agreement. It was extended for another three years, ending at the same time as the MFA phase-out on 31 December 2004. Along with the US-Cambodia Trade Agreement, an independent but complementary factory monitoring project, later named Better Factories Cambodia (BFC), was established in 2001 involving the International Labour Organization (ILO), the Cambodian government, GMAC and trade unions. Its initial purpose was to inform ongoing US quota allocation decisions. The ILO leads the project with funding from donor countries, the Cambodian government, GMAC and international brands. The ILO's Better Factories project is the first of its kind, later serving as a model for the ILO/IFC Better Work programme. In summary, throughout the US-Cambodia Trade Agreement era workers' rights have been discursively prominent, international buyers use it as a showcase for their ethical commitments, and the US the ILO and World Bank have promoted it as a new model of export-led development (Arnold and Toh 2010).

Underlying the BFC project is a set of mechanisms to initiate tripartite industrial relations systems in a country with no such history. A basic contention argued elsewhere (Arnold and Toh 2010) is that BFC played a defining role in Cambodia's garment sector in the early years. Much attention has been paid to the workplace specific initiatives that comprise BFC, particularly the impacts of monitoring on compliance and the role of international buyers, among other issues (see Brown, Dehejia and Robertson; Gregoratti and Miller 2009; Miller et al. 2007; Oka 2012; Polaski 2009, Chapter 10 in this volume). Lacking are studies on workers' and trade unions' bargaining power in the wider political economic context comprising capital, state, the labour market and workers' communities.[4] This is surprising since the then chief technical adviser of BFC contends that the heart of the programme is whether or not workers themselves have the collective voice and ability to defend their rights and fight labour rights violations (interview with Tuomo Poutiainen 16 September 2011). The large scale garment workers' strike in September 2010 is a defining moment in this process – namely workers and trade unions recognizing themselves as the active agents who must shape the outcomes of the structures and institutions built over the 2000s. The strike centred on wage and income, an issue that currently does not directly involve BFC and other international organizations, buyers, foreign governments, non-governmental organizations (NGOs) and other actors. Rather, trade unions and workers, GMAC and the government determined the outcome. In summary, BFC has been central in constructing a tripartite negotiating table in Cambodia, and the programme may continue to invite participants to it and facilitate. Yet it is 'embedded' stakeholders – trade unions and workers, manufacturers and state officials – who are primary in shaping the outcomes on an issue, wage, which matters most to a majority of workers. The following section examines the changing Cambodian state and its role in these processes.

The state: the Cambodian People's Party and protest space

Cambodia's political, economic and social context is unique. The 30 years of war from the late 1960s to late 1990s left Cambodia's state and society eviscerated and without a clear centre of political gravity or autonomous development agenda. Cambodia's specific history of social struggles, war and political disintegration is a key factor in understanding why the country entered the global economy from a position of weakness (Arnold and Toh 2010). Initially the US-Cambodia Trade Agreement was seen as a means for Cambodia's economy to be favourably inserted into the global economy and to overcome their past political and economic problems. Given that Cambodia's political economy was a relatively blank canvas, this provided opportunities in developmental experimentation for foreign governments, international organizations and NGOs who were granted wide latitude to operate (Arnold and Toh 2010). This created much dependency and lack of autonomy in the socio-economic development process, a situation few other countries have been willing to accept.

Following a coup in 1997, the Cambodian People's Party (CPP) came to power, led by Hun Sen. The CPP is a reformulated version of the Kampuchean People's Revolutionary Party who governed Cambodia during Viet Nam's occupation from 1979 to 1989. The CPP and Hun Sen have remained in power, winning disputed elections, in on-again, off-again coalitions with FUNCINPEC through the 2003 elections. In the 2008 National Assembly elections, the CPP further consolidated its position, winning 90 of 123 seats, securing Cambodia's position in the 'electoral authoritarianism' camp (Hughes 2007: 835). A common critique summed up by Hughes and Un (2011: 10), is that 'The Cambodian state is authoritarian, corrupt and based heavily upon neo-patrimonial institutions, whose survival and expansion represent key interests driving, and limiting, public policy'. The return to authoritarianism in Cambodia has gone virtually unchallenged in the international community. Springer (2009: 156) contends that 'the donor community views authoritarian action taken by Hun Sen and the ruling CPP as conducive to maintaining their own interests, although this remains unspoken and hidden behind rhetorical appeals for greater democracy'. This means Cambodia's government, or more accurately the CPP, has more power to govern economic policy and labour relations as it more deeply integrates with the global economy. Garment manufacturing is considered a key link.

As demonstrated more fully in the following section, Cambodia's garment industry is not deemed competitive in global markets due to high price, low quality and long lead times, in addition to other factors. A firm-centred global value chain approach includes the following policy recommendations (Staritz 2011: 131; see also USAID 2007):

1. Improving productivity, skills and capabilities at the firm level and developing from CMT (cut, make and trim) to FOB (freight on board) and full-package supplier;
2. Increasing backward linkages and reducing lead times;
3. Improving physical and bureaucratic infrastructure, particularly with regard to transport, logistics and customs, electricity and access to finance;
4. Diversifying end markets;
5. Increasing local involvement in the industry at the management and owner level; and
6. Increasing regional integration.

Hughes and Un (2011) counter such contentions, asserting that the failure of Cambodian state institutions to conform to liberal development models is a matter of political choice, rather than of technical incapacity. The Cambodian government has shown low effectiveness and coherence to solve governance issues including the high cost of electricity (four to five times higher than neighbouring Viet Nam and Thailand) and labour relations. Ear (2011: 91) argues that the industry has survived, in part, due to 'GMAC's capacity to get things done with the RCG [Royal Cambodian Government]', in particular the Ministry of Commerce. GMAC became the primary institution to solve collective problems associated with the industry while maintaining close relationships with the government.

The conflict-ridden industrial relations environment in Cambodia has been a longstanding frustration for GMAC. This, in combination with the government's directive to sponsor a range of unions as a means to undermine the independent and opposition-oriented unions (Arnold and Toh 2010; Ear 2011; Hughes 2007), means space for independent trade unions and other social movement organizations in Cambodia has been under constant pressure throughout the 2000s. Prime minister Hun Sen, in response to suggestions that he should be concerned about his position in light of the overthrow of the Tunisian dictator in 2011, gave a clear indication of his willingness to silence or put down those deemed to be challenging the government, both Cambodian and foreigners:[5] 'I not only weaken the opposition, I'm going to make them dead... and if anyone is strong enough to try to hold a demonstration, I will beat all those dogs and put them in a cage' (Adams 2012). If empowering the people entails the simultaneous disempowerment of those who currently occupy a privileged position in society, then Cambodian elites will try to impede any movement toward redistribution of power and wealth (Springer 2009).

In summary, if states remain the pivotal institutional apparatus that regulate the lives and politics of workers (Coe and Jordhus-Lier 2010) then the CPP is a powerful force delimiting the space for any kind of workers' agency. Furthermore, GMAC has a very strong hand with the government – given the lack of policy initiative on the part of the state; GMAC has enjoyed considerable legitimacy for having facilitated the alternating expansion and survival of the industry. As a result, tripartism in Cambodia is marked by the skewing of power toward the interests of the government and employers – a global pattern with unique characteristics in the case at hand. Either or both of these actors, as will be demonstrated, control or influence a majority of trade unions. As a result, rank and file workers and leaders of the few independent unions have limited opportunity to enter negotiations on an equal footing. It is not surprising that frustration builds and workers take to the streets to air their grievances and seek resolution to their demands for a decent wage.

Capital: the competitiveness in global markets

Cambodian exports are highly dependent on the garment industry, on a single end market and on a single low value-added product. Textile and garment exports have comprised roughly 90 per cent of Cambodia's annual export revenue throughout the 2000s to date. Roughly 70 per cent of these exports have been destined for the US market, making Cambodia the US's eighth largest apparel supplier in 2007 and 2010 measured by import value, accounting for just over 3 per cent of total US imports in both years. With the elimination of global quotas and more direct exposure to global competition Cambodia is relying less on its labour monitoring for a competitive advantage and increasingly on managing more generalized competitive factors such as productivity, reducing transaction time and cost, maintaining low wage, streamlining regulatory burdens and the like (Arnold and Toh 2010; EIC 2007; USAID 2007).

Despite widespread fears of the post-quota collapse of Cambodia's garment industry, exports expanded from 2005 through 2007. There appear to be three reasons for this (Arnold and Toh 2010). First, the imposition of US safeguard quotas against China in 2005, effective through 2008, aided Cambodian exports. Second, costs in coastal China have risen considerably, leading to capital relocation further inland China or to other countries including Cambodia. Third, Cambodia has been a part of a post-quota regional trend that has seen Asia-based suppliers increasing exports, while many suppliers in the Americas have seen significant declines (Textile Outlook International 2009). This trend points to the cost competitiveness of exporters in Asia. Overall, Cambodia has been a part of a growing concentration of US suppliers. In 2010, roughly 79 per cent of US textile and clothing imports in value terms came from the leading ten supplying countries, compared with only 54 per cent in 2003 (Textile Outlook International 2009). The trend toward sourcing in Asia has been volatile in Cambodia compared to Bangladesh and Viet Nam.

Despite this initial increase, Cambodia's exports declined sharply in 2008 and 2009 during the global financial crisis, rebounding in 2010 and 2011. For instance, garment exports dropped 21.6 per cent in the first nine months of 2009, with 70 factories closing and approximately 70,000 workers laid off (CIDS 2010). Interviews with GMAC, industrial zone operators, factory managers, industry observers, logistics providers and other stakeholders between 2006 and 2011 elicit the following explanations for both Cambodia's export downturn in 2008–2009 and the more general lack of competitiveness regularly discussed since the end of the global quotas in 2005, including: poor infrastructure, low workforce productivity, high frequency of strikes, high utility costs, high costs at customs and other 'administrative' costs (usually referring to corruption), a small domestic market (meaning nearly 100 per cent of garment products are exported), insufficient trade facilitation and access to credit, and 'dollarization' – where a rising US dollar hurts competitiveness. Cambodia's garment factories also have a structural disadvantage – 60 per cent are CMT or assembly factories, usually subsidiaries of corporations with operations in other countries (CIDS 2010). The majority of inputs are imported and factories do not generally engage in higher value-added operations in value chains. In Viet Nam and Bangladesh, by comparison, an increasing share of firms are developing more than CMT capabilities and are in charge of input sourcing and pattern making (Staritz 2011). Ken Loo of GMAC (interview 18 July 2011) noted that in 2010 and the first half of 2011 the price of fabric rose by 80–100 per cent, yet the price from buyers has only increased by 5–10 per cent. Combinations of these factors have led to higher prices in Cambodia.

Factories in Cambodia have several mechanisms at their disposal to lower costs. Subcontract factories are increasingly prevalent. Subcontractors are both registered and unregistered, yet are not GMAC members, and are outside of the Better Factories monitoring orbit. This creates challenges for implementing compliance programmes and trade union organizing efforts. The Coalition of Cambodian Apparel Workers Democratic Union (CCAWDU) estimates there are 200 unlicensed factories in the country (interview July 2011). Of those, 90 are subcontractors for licensed companies. One relates to the marked increase in the number and use of subcontract factories, both registered and unregistered, since 2005. CCAWDU has only one trade union in a subcontract factory, compared to about 60 members in registered GMAC factories. As competition with Viet Nam and Bangladesh becomes more robust it is expected that subcontract factories will play a larger role (interview with Tuomo Poutiainen, 16 September 2011). CCAWDU finds that these factories have led to a general degradation of working conditions throughout the industry. This is compounded by a lessened impact of BFC monitoring reports, since they are no longer tied to quota allocations to the US market. From the BFC perspective the subcontracting issue is central to overall standards in the industry and there has been a process to find ways and means to address subcontracting (interview with Poutiainen, 16 September 2011). It has resulted in new regulations, in 2011, concerning subcontracting with the aim to bring these units under better control, and possibly under BFC monitoring (interview with Poutiainen, 16 September 2011). Yet it is up to the Cambodian government to find a systematic way to deal with the subcontracting issue if they want to continue to link trade with labour rights.

Producers in Cambodia are generally lagging in price, speed and quality demands of buyers. Research shows that labour compliance is a consideration for buyers (Oka 2012), but it is clearly not enough to sustain an industry. In fact, the GMAC has made its position clear; that BFC adds little if any value to their competitiveness (interviews with Arnold 2006, 2008 and 2011). The 2008 crisis, when Cambodia's orders declined precipitously compared to competing Bangladesh with its lower paid workforce and lack of systematic compliance discourse or practice, confirms GMAC Secretary General Ken Loo's belief 'that indeed compliance or labour standards is icing on the cake – price is the cake' (Lazaro 2012). Manufacturers' commodity specialism, in this case producers at the lower end of the global garment industry, shape relations with workers. After nearly 20 years in global garments, the industry in Cambodia shows few signs of impending economic upgrading. To face workers' wage increase demands it is clearly in the interests of employers to maintain an atomized labour movement and a flexible labour force to react to market fluctuations. Fixed duration contracts are a strategy employers have adopted to manage both. The following section addresses this issue.

The labour market: fixed duration contracts

Under Cambodian labour law there are two main categories of employment contract: undetermined duration contracts (UDCs) and fixed duration contracts (FDCs). As their names suggest, UDCs are valid for an unlimited time, while FDCs are probationary or valid for a specific period of time, usually three to six months, and can legally extend indefinitely if workers and employers 'agree' on the terms. Kang et al. (2009: 18) find that FDC terms have shortened since the onset of the global recession in 2008, generally from six to three months. Since the end of global quotas one of the significant changes in factories has been the increased use of FDCs. According to an unpublished investigation conducted by the Workers' Rights Consortium in 2009 (Arnold and Toh 2010), only one of 60 factories surveyed exclusively used UDC workers, while most of the remaining 60 factories surveyed either exclusively use FDCs or since 2005 employ all new workers on FDCs. According to the most comprehensive study on FDCs to date: 'The widespread shift from UDCs to FDCs has resulted in tremendous worker insecurity, heightened antagonism between unions and factory management, and a threat to peaceful industrial relations' (Yale Law School 2011: 15). The main findings of the study contend that it is clear that the conversion to FDCs is not the result of voluntary and informed decisions made by workers, 'but a top-down decision imposed by employers, often through coercion, manipulation, or deceit ... [which] clearly facilitates employers' anti-union discrimination and suppression of free association' (Yale Law School 2011: 77).

FDCs have become a major, if not the defining, challenge for trade union organizing and workers' livelihoods. Employers are increasingly using FDC and daily labour as part of efforts to maintain or increase profits and, not coincidently, to avoid the demands of organized workers. Commenting on this trend to FDCs, Anne Ziebarth, legal adviser for Better Factories Cambodia, stated that this was 'troubling because it may indicate that they [employers] misunderstand the appropriate use of the different types of contracts, or that they are using FDCs to undermine workers' employment security' (BFC 2006b).

The ILO confirms that employers prefer workers on FDCs to those on UDCs 'because they believe that it is easier to terminate workers' (BFC 2006b: 1). Employers are under no obligation to renew a flexible worker's contract after it has expired, but they are prohibited from firing FDC workers for illegitimate reasons, including anti-union discrimination. The expansion of FDCs is a major factor in the increase in plant level strikes in Cambodia since 2005 (BFC 2006b). Increasing frequency of strikes is considered a detriment to Cambodia's 'competitiveness' vis-à-vis China and Viet Nam. These claims overlook fairly widespread and often large scale strikes in China and Viet Nam over the past decade (Lee 2006; Lee 2007). And, as demonstrated in the case study in the workers' agency section, recent strikes in Cambodia have not led to widespread reduction in orders.

The increasing prevalence of FDCs signifies a break from the decent work principles of the BFC programme, including job security, benefits and rights to freedom of association. In particular, an objective of the ILO programme is to encourage bureaucratic solutions attained by professional negotiators to decrease the likelihood of disputes disrupting industrial production (Hughes 2007). Use of FDC workers appears to be a strategy by employers to compete in global markets, yet it produces discontent among workers and independent trade unions who increasingly take to the streets to make demands not met through institutionalized channels. Most importantly, FDCs negatively affect workers' livelihoods (Yale Law School 2011). This is critical anytime, but particularly when young women workers are expected to send home remittances despite high inflation and stagnant wages. It is an issue that generates discontent and anxiety among workers (Arnold and Toh 2010).

Community: the proliferation of trade unions

From the late 1990s, garment workers became part of a new social force emerging from the process of economic change in Cambodia, as it shifted from an almost exclusively agrarian to industrializing economy. Since the initiation of the US-Cambodia Trade Agreement and Better Factories there has been a rapid proliferation of trade unions in

Cambodia's garment industry. While GMAC counted 892 trade unions in 270 GMAC-member factories in mid-2006, the ILO statistics count 440 active unions in 314 factories in 2008. The ILO figure equates to 1.4 unions per factory, up to 1.7 in 2011, with total membership accounting for roughly 60 per cent of the workers in the industry. Roughly 84 per cent of factories have at least one union. In 2008 there were 24 union federations in the country, and by 2011 it increased to 29. An overwhelming majority of federations are aligned with the ruling CPP (Arnold and Toh 2010). Sixty per cent union membership is usually considered to demonstrate high levels of worker representation. However, in the Cambodia context the effectiveness of having so many unions is questionable, leaving many unions weak, under-funded, competing with one another and subject to corruption and political interference (Arnold and Toh 2010).

From the perspective of the handful of independent federations, the CCAWDU the most prominent from the mid-2000s to the present, their activities promoting workers' rights are inhibited by four factors:

1. Government (which often directly or indirectly supports 'friendly' unions);
2. Employers and the GMAC;
3. The numerous pro-capital unions that contend with independent and other unions for representation in factories; and
4. The mafia-unions that extort money from both employers and members.

There are several implications of this situation. First, despite high union membership, basic rights are often neglected or abused. For example, overtime work beyond legal limits and occupational health and safety issues have not been sufficiently addressed at the industry-wide level (interview with Tuomo Poutiainen 16 September 2011). Furthermore, unions claim that freedom of association is regularly denied and the increasing prevalence of FDC contracts means workers are reluctant or even unable to promote their rights whether individually or as union members. Second, and related, the high number of federations and plant level unions makes it difficult for 'genuine' unions to promote the rights of their members. Competition and conflict among the unions and the national federations limits their effectiveness. Third, the proliferation of unions and federations is associated with the rise of unions as businesses. Running a union can be lucrative, as corrupt unions can seek kickbacks from employers and 'fundraise' in other ways. Fourth, an atomized and corrupted labour movement loses political economic effectiveness. Such unions and federations have not posed a serious challenge to employers or to the ruling CPP's policies; an important consideration as textile and garments is the most highly organized sector in Cambodia. CCAWDU has claimed that the CPP is also encouraging the formation of friendly federations to prove that opposition/independent federations are a minority and do not represent the workers. Anti-union discrimination and suppression take other forms in Cambodia. Managers have sometimes fired workers who either openly sympathize with or join trade unions (Stanford Law School and Worker Rights Consortium 2013). Employers have also used legal threats against unions (Gregoratti and Miller 2009), as discussed in detail further. The past decade has seen recurring incidents of violence – from leaders of federations, to factory-level unionists, to rank and file workers (Arnold and Toh 2010; Stanford Law School and Worker Rights Consortium 2013). The most serious form of intimidation and suppression has been the murders of three trade union leaders since 2004, including union leader Chea Vichea, a once prominent opposition political figure.

If trade unions are meant to be a meaningful component of workers' lives that promote and protect their interests in the workplace and the community, then this section has not offered a flattering picture of the situation in Cambodia. Despite high levels of unionization, trade unions largely struggle to maintain representation at the factory level. Union leaders are also overwhelmed with court cases, arbitration and disputes, in addition to managing participation in a host of trainings, workshops and related matters. This is a critical factor preventing them from dedicating time and resources toward becoming more prominent social forces representing the interests of their members both inside and outside of the factory. In cases where union activists become prominent politically they are more susceptible to violence or even death, as was the case with Chea Vichea. How then to explain a mass mobilization in 2010 that resulted in a relatively peaceful and meaningful expression of workers' interests in factories and streets?

Workers' agency: a strike case study

Cambodia's garment workers are under increasing pressure to produce more for stagnant or declining wages, while families continue to rely on their remittances. A Cambodia Institute of Development Study (CIDS) report analysed the costs of basic living expenses for garment workers and found that a reasonable living wage (i.e. the standard required by

Article 104 of the Labour Law 1997) is US$93 per month (CIDS 2009a). A subsequent study found that the absolute minimum wage on which a garment worker could afford to live was between US$72 and $75 per month (CIDS 2009b). The key difference between a 'living wage' and a 'minimum wage' in the two studies was the ability to make small personal savings. Evidence suggests that workers earning below US$72–75 compromise basic nutrition. One worker said, 'I spend about 5,000 Riels ($1.19) per day on food. It is enough based on my salary, but not enough to keep me healthy. I can't afford to buy foods with high nutrition, like fruits' (CIDS 2009b: 29). Half of the workers surveyed who were earning around US$59 per month had to overcome this insufficient wage by reducing spending on food and healthcare, while others resorted to borrowing from external sources such as friends, family and private money lenders (CIDS 2010).

Pay is low in Cambodia's garment industry and is considered one of the – if not the – primary competitive advantages. Minimum wages have risen from US$45 per month for 'regular' workers (non-probationary) in 2000, to US$50 in 2006 and US$56 in 2008 after a living allowance was allocated due to 25 per cent inflation that year. After the 2006 minimum wage negotiations the next round of negotiations were set for 2010. On 8 February 2010 the Labour Advisory Committee convened a meeting to discuss the scheduled negotiations. The Labour Advisory Committee is a tripartite institution that, through late 2010, comprised seven representatives from trade unions federations representing garment workers, seven representatives from GMAC representing employers and fourteen representatives from the RCG. Despite the February decision that negotiations would take place, no meeting of the Labour Advisory Committee had been scheduled by the end of April 2010, nor had GMAC agreed to the bilateral wage negotiations favoured by many unions. In response to the perceived lack of action, various trade unions began planning rallies and demonstrations calling for negotiations and to build support for a wage increase.

By the end of May, the CCAWDU (member of the Cambodian Labour Confederation) began to publicly threaten strikes in response to the lack of action on the minimum wage negotiations. On 23 June, at an ILO-sponsored meeting in Phnom Penh, the Labour Advisory Committee group of unions agreed on a common position for negotiations, including the US$93 'living wage' derived from the aforementioned study. On 24 June, the Ministry of Labour and the Ministry of Social Affairs released a joint press statement that forwarded a recommendation from the prime minister Hun Sen that the minimum wage increase by US$5 per month, and that the existing US$6 cost of living allowance from 2008 be included in the revised minimum wage. This would mean a minimum wage of US$61 for unlimited duration workers, and US$56 for fixed duration workers. The government offered no rationale or study to support this recommendation or how it adheres to Articles 104 and 107 of the Labour Law.

A Labour Advisory Committee meeting was subsequently scheduled for early July. At this meeting, certain unions argued for the US$93 wage in accordance with their earlier agreement. GMAC reportedly claimed that this increase was too large due to their difficulties associated with the global economic crisis, yet did not offer an alternative proposal. According to the BFC then chief technical adviser, GMAC did not take the wage negotiation process 'as seriously as they perhaps should have' (interview with Tuomo Poutiainen, 16 September 2011). GMAC did not provide a study arriving at a sufficient wage that meets workers' needs, and this was a major factor in the ensuing strike. After less than two hours of negotiations the chairman of the Labour Advisory Committee concluded that there was no agreement and called a vote on the government's proposal (a US$5 increase and inclusion of the existing cost of living allowance). Of 28 in attendance only three voted against the proposal: one from government and two union representatives, Ath Thorn, president of the CCAWDU, and Morm Nhim, president of the Cambodian National Confederation and National Independent Federation Textile Union of Cambodia. Three government representatives were not present. The Ministry of Labour later confirmed that the new minimum wage would come into force on 1 October 2010. The two unions who voted against the proposal decided to continue campaigning for a higher wage, seeking to press GMAC and the government into re-opening negotiations.

The unions' opposition to the Labour Advisory Committee decision was based on the contention that the process had no transparency, lacked any logic or rationale for the US$5 increase and was insufficient to meet the basic survival much less living wage needs of workers. Another argument was that the Labour Advisory Committee appeared tripartite in form, but not in practice. In other words, the government recommended a wage increase that would satisfy the GMAC – the primary reason they had sought bipartite negotiations from the beginning of the process. Unions initially held a common platform, but this broke down, according to CCAWDU, due to co-optation to employers and government interests, rather than workers' interests. Subsequent requests to re-open discussions were denied by both GMAC and the government. The unions continued to organize members to push for further negotiations. In response, government officials began to publicly threaten legal sanctions against union leaders who continued to express opposition to the decision by organizing worker demonstrations. For instance, on 23 July, the secretary of state of the Ministry of Labour, Mr Oum Mean, stated publicly that, 'Ath Thorn would face criminal offences if he still fought against the decision of the Ministry.' Council of Ministers official Pa Angtoni said that he would investigate the possibility of charging union leaders with incitement, for which they could, if convicted, receive prison sentences of between one and five years.

On 18 August the eventual strike leaders filed a letter with their demands, to reconvene wage and benefits negotiations, to the GMAC and the Ministry of Labour and Vocational Training. This, unions claim, met Labour Law requirements for strike notification, set for 13–18 September. Despite a frenzy of municipal level court rulings against groups of striking workers at individual factories, no ruling disputes the legality of steps taken by the unions. The unions' demands were not met and the strike began on 13 September. According to union figures, on the first day of the strike 68,380 workers from 56 factories participated. By the third day, 201,770 workers from over 100 factories were participating. This would represent over two-thirds of all workers in garments and footwear. GMAC later claimed that at its peak the strike included only 10 per cent of total workers (interview with Ken Loo, 11 July 2011), or some 30,000 workers. It is safe to estimate a figure between the two – or around one-third of the total textile and garment workforce. This would make it the largest industrial action in Cambodia's history. Regardless of the numbers, the magnitude appeared to have been unexpected, particularly for GMAC, but also for the unions leading the strike.

During the four-day strike both government and GMAC members' reactions were aggressive. CCAWDU received three different court orders (dated 15, 16 and 17 September 2010) from the Kandal Provincial Court and from the Phnom Penh Municipal Court. The warrants stated that all striking workers must return to work within 48 hours. The warrants also stated that certain workers, named in the warrants, were forbidden from returning to work and the factory area until the case was resolved. No specific date was mentioned, nor was the nature of the resolution. Furthermore, in the warrant the CCAWDU president, Ath Thorn, was ordered to stop 'threatening and leading workers' in an 'illegal strike'. Observers concluded this was an unjustified crackdown on a legal strike action (Acils et al. 2010). The director of LICADHO, a prominent human rights organization, contended, 'This misuse of the court to punish union activism is a blatant violation of the Labour Law' (Acils et al. 2010). In summary, the situation over the first three days escalated. More workers participated in the strike with each day. Provincial and municipal courts issued irrational rulings that contradicted the country's laws. Furthermore, according to an anonymous source, there were concerns that the strike was affecting the Cambodian economy. Perhaps more importantly, voices emerged suggesting that it was an act of dissent against the ruling CPP. In the Cambodia context, as already demonstrated, this is a serious matter.

Finally, the night of the third day of the strike, the prime minister Hun Sen called the strike leader personally via cell phone and asked that the strike end. The following day, on 16 September, the minister of social affairs, youth and rehabilitation and the president of the Industrial Relations Working Group requested that trade unions and other stakeholders enter discussion on 27 September 2010 to seek a resolution to the strike. On the same day the unions convened a meeting with strike leaders and they agreed to end it, with all members to return to work the following day.

On 17 September, at least 14 different factories posted notices in front of the factory gate stating that certain strike leaders were not allowed to return to work. In total, according to CCAWDU, nearly 900 striking workers were suspended or sacked during the strike. This hindered and may have been intended to eliminate CCAWDU's factory level leaders' activities in the workplace in the weeks and months following the strike. In July 2011, 141 workers and activists from 13 factories were still seeking reinstatement. This is significant since in the days after the strike Hun Sen's and subsequent government requests were issued stating that employers should reinstate all workers and drop court cases. GMAC demanded a 'public apology' for the strike before all workers would be reinstated. CCAWDU refused, from their perspective they had done nothing to contravene the law at any point in the process. GMAC later claimed that the contracts of these workers had expired.

Over the course of ensuing months a reformulated Labour Advisory Committee met, this time including five representatives each from government, employers and unions. All sides agreed that minimum wage negotiations would not reconvene until 2014, GMAC's primary position. Instead workers' benefit packages and bonuses were negotiated and resulted in: attendance bonuses increased by US$2 per month; increased seniority pay; and increased meal allowance from US$0.25 to

$0.50 per day for overtime, among other points. CCAWDU estimates the combined increase equals roughly $10 per worker per month. GMAC puts the increase at about US$7 per worker per month (interview with Ken Loo 11 July 2011), in addition to the minimum wage hike set by the Labour Advisory Committee in July. GMAC estimates that the average wage in 2011, including overtime, allowances and benefits, is US$94 per month. The ILO figure is US$87–94 per month.[6]

Prior to the strike the leaders expressed concern over the potential of physical violence and legal action against strike leaders and workers. The strike provided a brief sense of optimism that the government was willing to mediate capital–labour relations. For unions it was a risky move – only two of 29 garment federations mounting a strike in the midst of a global economic crisis, knowing they would face a hostile reaction from employers. On the one hand, it made independent unions a 'part of the process' from which they had felt excluded. On the other hand, GMAC now sees them as the proactive unions linked to global solidarity and support movements. GMAC seeks to mitigate this influence with workers and other unions, an integral aspect of capital–labour relations.[7] For their part, workers challenged their place in the hierarchy of garment GPNs in Cambodia – a significant step for rural women in a patriarchal society and sector dominated by powerful foreign investors with the backing of Cambodia's CPP.

Conclusion

Cambodia's garment industry is a paradox – it is characterized by a highly confrontational industrial relations environment, but maintains a systematic workplace monitoring programme. It is promoted as a 'showcase' for ethical production (Polaski 2009), yet tripartite stakeholders consist of an 'electoral authoritarian' regime (Hughes 2007), an employers' association that is openly hostile to both the BFC programme and independent trade unions, and a remarkably high number of trade union federations that preclude attempts to build a labour movement. Furthermore, the proliferation of fixed duration contracts facilitates employers' anti-union discrimination and suppression of freedom of association. This chapter has focused on Cambodian garment workers' potential for action in relation to these particular formations of capital, the state, the community and the labour market. The nationwide 2010 garment workers' strike demonstrated that Cambodian workers have agency, albeit limited, to shape wage negotiations. Indeed, only a small proportion the workers' demands were met, reflecting workers' precarious position vis-à-vis capital, the state and even a majority of trade unions, forces that continue to tightly delineate workers' agency. Viewed in isolation, the strike could be considered a failure. Yet, in an authoritarian and often violent political context, it is a major step for independent unions and workers to make their collective voice heard. Increasing labour unrest and regular media coverage of workers' hardships has brought workers' concerns to the centre of national political debate. A pillar of the opposition CNRP's National Assembly campaign in 2013 elections was the promise of a $150 per month if elected. In response, amidst widespread discontent that the election results were fraudulent, the ruling CPP confirmed garment wage negotiations would take place in 2014. Workers are clearly an important voting bloc, yet a unified strategy to ensure implementation of a living wage is at beginning phases.

In the absence of a living wage, workers face may difficulties. From mid-2010 throughout 2012, mass faintings occurred in numerous garment and shoe factories in Cambodia. The cause has not yet been determined, but a contributing factor may be malnourishment and excessive working hours (interview with Tuomo Poutiainen, 16 September 2011). If that is the case, the overall wellbeing of garment workers in Cambodia is being sacrificed for economic gain. This brings the lack of social protections and poverty to the foreground. Both BFC's and trade unions' efforts to address these and other social concerns are oriented around tripartism and a Fordist-inspired workplace specific mode of engaging workers. Yet the informalized nature of work in Cambodia's garment factories restricts possibilities for enduring forms of collective action. This, in turn, restricts the full potential of workers' collective labour power. In a context where young women are expected to keep sending remittances to families and communities despite inflation, irregular work and other factors beyond their control, new forms of organizing and social protections are needed – that do not privilege the workplace as the unit through which social change is channelled. Thus, there is a need to better understand and conceptualize how labour is situated in community and social relations in Cambodia and other countries, and the ways in which workers and their communities shape labour politics, local development processes and global production networks.

Notes

1. This chapter is based on fieldwork in Cambodia from April to June 2006, December 2008 to May 2009, August to September 2010, December 2010

and May to July 2011, and interviews with stakeholders in Cambodia via skype in September 2011. Certain sections draw on Arnold and Toh 2010. Qualitative methods were primary, including participant observation with the CCAWDU, interviews and focus group discussions with stakeholders including GMAC, government officials, international and multilateral organizations, NGOs, factory and industrial estate managers, development consultants and researchers, and other labour, industry and development analysts. This is supplemented with descriptive data analysis. Among the long list of people to thank, CCAWDU are foremost, especially Kong Athit for his thoughtful insights and analysis.

2. Food sellers represent almost 40 per cent of indirect jobs, housing 25 per cent and transportation 5 per cent. Another 30 per cent include small traders, clothing shops and other supporting businesses (Economic Institute of Cambodia 2007: 16).
3. According to the Sam Rainsy Party leader Mu Sochua (Sochua and Wikstrom 2012), roughly 420,000 Cambodians have been affected by evictions since 2003.
4. See Arnold and Toh (2010) for a garment specific study, and Hughes (2007, 2011) for analysis of Cambodia's political economy and implications for labour.
5. For example, opposition leader Sam Rainsy was in exile from 2008 to 2013. In 2004 one of the opposition's staunchest supporters, Chea Vichea, president of the Free Trade Union of Workers of the Kingdom of Cambodia (FTU), was murdered in an open market; the killers have never been found. Christophe Peschoux, the former head of the local UN Office of the High Commissioner for Human Rights, was 'reassigned' in 2011 after his long running criticism on issues such as land grabbing and crackdowns on government critics.
6. On 11 July 2012 the Labour Advisory Committee agreed on a further increase: (1) Increase regular attendance bonus from US$7 to US$10 per month, and (2) US7$ per month is provided by garment and footwear employers for transportation and accommodation.
7. When asked 'what are the implications of the strike in terms of labour relations?', the secretary general of GMAC said, 'GMAC now refuses to work with CCAWDU. That's the implication' (interview with Ken Loo, 11 July 2011).

10

Factory Decisions to Become Noncompliant with Labour Standards: Evidence from Better Factories Cambodia

Drusilla Brown, Rajeev Dehejia and Raymond Robertson

Introduction

Efforts to improve working conditions in developing countries participating in global supply chains have generated a large and growing literature (Elliott and Freeman 2003; Harrison and Scorse 2010). Much of this literature is motivated by the common perception of very poor conditions in developing country factories (the term 'sweatshop' appears often in this literature). As a result, the usual approach is to focus on the debate around policies (or potential policies) to improve working conditions. One key issue in the debate centres on the optimality of such improvements (Brown et al. 2011a). While such improvements may be beneficial to workers, the effects on the factory as a viable economic unit are less clear. The conventional wisdom in economics suggests that if factories are optimizing and efficient, imposing additional constraints from the outside, including from other actors in the GVC, pushes them from their privately-determined optimal practices.[1] Poor conditions are the result of optimizing behaviour of firms, and attempts to improve working conditions from the outside threaten factories and, therefore, the jobs of the very workers that the policies were intended to help. Some have even suggested that 'sweatshop' conditions are a necessary stage of economic development.

Recent data from Cambodia, however, suggest that compliance with national law and ILO core labour standards may not be as low as commonly believed. Since 2001, the Better Factories Cambodia programme has been monitoring detailed working conditions in exporting garment factories producing for global buyers. Overall compliance rates are over 50 per cent, with the lowest aggregated compliance rates reported by Ang et al. (forthcoming) being over 60 per cent. They report an overall average of compliance rates over all visits and all categories of 87 per cent and an average of first visit compliance rates across all factories and all categories of 82.4 per cent. These relatively high compliance rates may suggest that compliance in many of the areas monitored by the programme is optimal for the factories.

Given these relatively high compliance rates, it is possible that we may learn a great deal about factory decisions by analysing the decision to become *noncompliant*. Identifying factors affecting this decision may provide information that would help identify the kinds of changes that are beneficial, the factors that may affect the sustainability of such improvements and the balance of costs and benefits for various areas of working conditions.

This chapter analyses the decision to become noncompliant in the context of Cambodian apparel factories. The 1999 US-Cambodia Trade Agreement created the incentive for factories to improve working conditions by linking such improvements to increased access to the US market (Berik and van der Meulen Rogers 2010; Polaski 2004). Such market access captured the attention of Cambodian producers because at that time apparel trade was restricted by the Multi-Fiber Arrangement (MFA) regime. The ILO Better Factories Cambodia (BFC) programme was created to monitor and help factories improve working conditions. The monitoring reports were used by the US government to determine Cambodia's apparel export quota allocation.

The BFC programme has captured a great deal of attention in academic literature. Combining interviews, observations and BFC synthesis reports, Shea et al. (2010) document sustained improvements in working conditions in Cambodia. Others, such as Polaski (2006), Adler and Woolcock (2010) and Miller et al. (2009) also document progress linked to, and concerns about, the BFC programme and generally agree that the programme has made positive contributions towards improvements in working conditions in Cambodia.

The market incentive linked to the MFA and the BFC programme was probably not the only factor affecting working conditions. Several papers that analyse the effects of the BFC programme consider other factors that might also have been at work. In particular, Oka (2010a,b) and Robertson et al. (2011) show that the presence of a reputation-sensitive buyer and the policy of public disclosure of non-compliance through the BFC programme both increased the likelihood of compliance. Some of these factors changed during the 2001–2011 period.

Although the MFA ended on 1 January 2005, Beresford (2009) in particular finds that, overall, working conditions did not fall in response to an increasingly competitive environment. Ang et al. (forthcoming) find that ending the programme of public disclosure reduced the rate of compliance, especially between the first and second visit. Another challenge emerged with the global financial crisis in late 2008. Exports fell and recovered, but the effects on compliance, much less the decision to become noncompliant, have not been examined.

To analyse the decision to become noncompliant, this chapter uses factory-level data from the BFC programme. The dataset has several characteristics that make it ideal for analysing the decision to become noncompliant. First, the data track individual factories over time so that individual fixed effects can be included in the empirical analysis. Second, the data include over 400 measures of individual compliance points allowing us to identify the effects of various factors on both broad and narrowly defined areas of compliance. Third, the data span several important changes in environment that could affect the decision to become noncompliant, allowing us to separately identify the effects of each change.

We contribute three main results to the literature. Our first main result is that the rates of becoming noncompliant are overall quite low. Even when the market access incentive is diminished, the rates of the decision to become noncompliant do not increase dramatically. Second, consistent with capital investment literature, adjustment costs matter. Costly investments are less likely to be reversed. Third, factors that affect compliance also, not surprisingly, affect the decision to become noncompliant. In particular, public pressure also matters: we find a significant break in behaviour following the BFC policy change in 2006 related to public disclosure of factory-level working conditions. Together these results do not seem to support the hypothesis that improvements are harmful to factories. On the contrary, taken together the results seem consistent with the hypothesis that such improvements help factories.

Data description, summary statistics and factor analysis

The BFC programme was established by the ILO in 2001. Since the programme is becoming increasingly well-known, readers are referred elsewhere for a detailed description of the programme.[2] It combines monitoring, remediation and training with the goal of improving working conditions in apparel-exporting factories. Monitors observe working conditions in all apparel-exporting factories during unannounced visits.

To avoid monitor bias, each monitoring team contains at least two people, and the same team rarely assesses the same factory twice. After the factory's second BFC visit, BFC publishes the firm's name and progress on improving working conditions in an annual synthesis report, which they share with the factories' buyers. Much of the recent literature that focuses on the BFC programme use data contained in these publically available synthesis reports (e.g. Beresford 2009; Shea et al. 2010).

National law mandated universal participation, reaching 119 factories in the original 2001–2002 wave of visits. In the following three years, however, monitors focused on following up on previously noncompliant items rather than comprehensively assessing factories against the full checklist. As a result, data are unavailable for this three-year period. The launch of the improved Information Management System survey in December 2005 marks the beginning of the next wave of documented visits in which monitors have visited each factory once every ten months on average.

Summary statistics

Rather than aggregate synthesis reports, this chapter uses factory-level monitoring reports generated by the BFC programme. Table 10.1 shows the number of factories by visit by year for the 2001–2011 period. New firms entering each year (with a first visit) and existing firms accumulating visits generate the table's upper triangular structure. The total of 2,113 observations is the product of 446 individual factories times each factory's number of individual visits (the maximum number of visits observed for any factory is ten). Visits typically fall about ten months apart, but the time between visits varies widely. National ownership also varies. The vast majority of the sample (93.7 per cent) is foreign owned, with 42 per cent owned by China, Hong Kong Special Administrative Region and Macau Special Administrative Region, 23.3 per cent owned by Taiwan and less than 3 per cent owned by Western countries.

Table 10.1 also reveals significant attrition in the data. While there are a total of 446 factories with an initial visit, there are only 241 with a fifth visit. Much of the lack of fifth visit observations comes from the fact that the second 'wave' is relatively large. Since tracking factories over time is important, care is taken to identify factories that have actually closed rather than simply changed names. An official list maintained by the BFC programme of confirmed closings is combined and the addresses of the factories are compared over time. Fewer than five have the same address with distinct names (the same factory identifier is used for these observations). If a factory closes and then re-opens at

Table 10.1 Factory visits by year

Visit	Visit year									Total
	2001	2002	2005	2006	2007	2008	2009	2010	2011	
1	85	34	7	188	30	37	27	20	18	446
2	0	0	18	122	136	34	28	16	6	360
3	0	0	0	48	186	33	24	27	5	323
4	0	0	0	0	80	152	27	20	11	290
5	0	0	0	0	11	112	82	24	12	241
6	0	0	0	0	0	38	102	42	12	194
7	0	0	0	0	0	0	52	75	20	147
8	0	0	0	0	0	0	11	43	28	82
9	0	0	0	0	0	0	0	13	12	25
10	0	0	0	0	0	0	0	3	2	5
Total	85	34	25	358	443	406	353	283	126	2,113

Notes: Data are missing for 2003–2004 because BFC monitors concentrated on previously-identified issues rather than completing a full evaluation. See text for details.

another location with a different name and different ownership (e.g. Macau Special Administrative Region may have a factory that closes and passes its business to a firm owned by China), these are treated as separate factories.

Working conditions are evaluated using 405 individual questions, such as 'Has management appointed a liaison officer?'; 'Are women paid their maternity leave benefits either before or during leave?'; and 'Does management keep an up-to-date list showing each worker's schedule for weekly time off?' These questions are then compared to domestic law and international standards and coded into binary variables that indicate compliance. Of these 405 questions, 62 show no variation across both factory and visit. These questions are dropped from the analysis. The remaining questions are analysed at three levels of aggregation: six factors that are identified through factor analysis, 31 aggregate compliance groups that roughly conform to groups commonly used by the ILO, and individual questions.

In practice, the measures of the decision to become noncompliant also vary by aggregation level. At the individual question level, the decision to become noncompliant is measured using a binary variable equal to one for factory-question observations that have become noncompliant (and zero otherwise). For aggregate levels, the measure of the decision to become noncompliant is calculated as the change in the average value across all questions within each group for each factory. This approach gives a continuous measure of compliance that allows assessment of the magnitude of changes as well as the direction. Therefore, it is analysed as either a negative change in the compliance average or with a binary indicator for the cases in which the change in the aggregate measure at the factory level was less than zero. All changes are based on the difference between the current and the previous visit unless otherwise indicated.

Table 10.2 contains the average of the binary measures of reduced compliance for the 31 aggregate compliance groups for 2006, 2008 and 2010. The changes vary considerably across groups. The core labour standards have very little reduction in compliance. Core labour standards such as forced labour and child labour are *zero-tolerance* points from the perspective of reputation-sensitive buyers. These are highly sensitive issues for buyers that are extremely likely to result in the buyer ending the relationship with the factory.

Areas related to industrial relations such as shop stewards, unions and strikes have some of the lowest rates of reduced compliance. This may be because these changes are relatively costless to the factory to

Table 10.2 Regression summary statistics for 31 compliance groups

Group	2006	2008	2010
Forced labour	0.357	0.005	0.000
Sexual harassment	0.065	0.051	0.015
Strikes	0.000	0.000	0.000
Accidents/illnesses compensation	0.224	0.024	0.038
Unions	0.006	0.008	0.004
Discrimination	0.077	0.141	0.160
Internal regulations	0.047	0.060	0.042
Child labour	0.071	0.035	0.027
Collective agreements	0.200	0.244	0.262
Liaison officer	0.394	0.301	0.297
Disputes	0.371	0.252	0.247
OSH assessment/recording/reporting	0.377	0.320	0.224
Machine safety	0.229	0.157	0.152
Shop stewards	0.006	0.000	0.000
Discipline	0.135	0.089	0.049
Emergency preparedness	0.076	0.024	0.046
Maternity benefits	0.207	0.236	0.335
Drinking water	0.231	0.146	0.110
Information about wages	0.331	0.352	0.338
Regular hours/weekly rest	0.408	0.211	0.156
Food	0.254	0.266	0.278
Contracts/hiring	0.255	0.190	0.274
Holidays/annual/special leave	0.284	0.214	0.300
Health/first aid	0.089	0.122	0.125
Sanitation	0.408	0.325	0.319
Workplace operations	0.284	0.206	0.163
Termination	0.341	0.306	0.285
Overtime	0.147	0.257	0.236
Payment of wages	0.067	0.030	0.000
Chemicals	0.396	0.263	0.202
Temperature/ventilation/noise/light	0.148	0.230	0.198

Notes: OSH = occupational safety and health. These measures are calculated by taking the average of the binary compliance indicators across individual questions within each group and across all existing factories in the sample for each year.

implement and, once these changes are put in place, it would take a deliberate and conspicuous effort to remove them (going from having a shop steward to not having a shop steward, for example). It is important to note that areas such as 'strikes' and 'unions' do not necessarily imply that an active union that engages in strikes is present in the factory. These areas are based on questions that are designed to capture compliance with national labour law with respect to these areas. Note that the areas of 'collective agreements' and 'disputes' have relatively high rates. These areas capture potentially contentious areas that may easily change through time depending on changes within the factory.

Others, such as 'holidays/annual/special leave', 'termination' and 'maternity benefits', also have relatively high rates. These illustrate areas that do not necessarily represent an investment on the part of the factory. Violations in these areas might also be the result of idiosyncratic situations that may not have emerged or tested the factories in earlier surveys.

For most areas, there is no clear trend overall in reduced compliance rates over time. They increase across years for some, fall across years for others and exhibit other patterns as well. One area that seems to show consistent improvement over time is the rates of reduced compliance in 'payment of wages' and 'regular hours/weekly rest'. These are the areas that seem to be most closely linked to the 'efficiency wage' literature that suggests workers who are paid and get adequate rest might be more productive and, therefore, offer direct benefits to factories through compliance. This is an example of the possibility that individual groups' patterns are driven by underlying factors. If underlying factors are driving compliance decisions of compliance questions to change in similar ways, then grouping questions that move together help present a more intuitive and compact set of results. Grouping questions, however, can be subjective. To minimize the subjectivity of our groupings, we apply factor analysis. Factor analysis is an empirical technique that identifies common movements in underlying data to help identify groups of results whose common movements may be driven by common underlying factors.

Factor analysis

Factor analysis helps identify a few common factors that may explain common changes in individual categories. The groupings are admittedly subjective in factor analysis and, therefore, the steps are explained carefully.

The core labour standards of child labour, forced labour and discrimination start with generally high compliance and vary little, so they are grouped. Then an orthogonal rotation is applied to the results generated by applying the principle-factor method to the remaining 28 of the 31 compliance categories.[3] The resulting matrix identifies nine possible factors, but none of the maximum values appear in factors five and eight, so attention is focused on the remaining factors. Although involving a combination of subjective judgement and interpretation, it appears that the emerging pattern enables the 31 categories to be sorted into the five additional factors shown in Table 10.3.

Table 10.3 Groupings resulting from factor analysis

Factor 1: Communication and workplace systems		**Factor 4: Compensation**	
6	Shop stewards	10	Payment of wages
7	Liaison officer	11	Contracts/hiring
23	Workplace operations	16	Internal regulations
		29	Accidents/illnesses
		30	Holidays/annual/special leave
		31	Maternity benefits
Factor 2: Occupational safety and health		**Factor 5: Unions**	
17	Health/first aid	4	Collective agreements
19	Temperature/ventilation	5	Strikes
20	Drinking water	8	Unions
21	Sanitation	14	Sexual harassment
22	Food	15	Disputes
24	OSH assessment/recording		
25	Chemicals		
26	Emergency preparedness		
Factor 3: Modern HR Practices		**Factor 6: Core labour standards**	
9	Information about wages	1	Child labour
12	Termination	2	Discrimination
13	Discipline	3	Forced labour
27	Overtime		
28	Regular hours/weekly rest		

Table 10.4 contains the summary reduced compliance measures for the six factors identified in Table 10.3. Some of the compliance groups in Table 10.4 fall (unions and core labour standards). Others rise and then level off (compensation). The rest rise over time. The rates are also higher. The rates start relatively low in 2006, but increase in 2010. While always below 50 per cent, some of the rates, such as 'modern HR practices' in 2010, seem very close to 50 per cent. The core labour standards remain relatively low, possibly suggesting that these are considered important to buyers or have stronger consequences associated with non-compliance.

One of the concerns with the aggregate measures presented in Tables 10.2 and 10.4 is that they include composition effects. The entrance of factories that are more likely to regress would increase average reduced compliance rates. It seems likely that both of these results are due to aggregation and the definition of the decision to become noncompliant

Table 10.4 Aggregate regression rates for underlying factors

Factor	2006	2008	2010
Communication and workplace systems	0.095	0.278	0.332
Occupational safety and health	0.142	0.451	0.488
Modern HR practices	0.159	0.421	0.495
Compensation	0.115	0.392	0.392
Unions	0.142	0.116	0.085
Core labour standards	0.176	0.052	0.014

Notes: The individual factors are comprised of various groups as described in Table 10.3.

that we use. We control for this by using question level measures of the decision to reduce compliance in the analysis.

Empirical analysis

In this section there are two approaches taken. First a Chow-like test is applied to evaluate the possibility of a structural break in the pattern of the decision to become noncompliant as a result of a policy change in the middle of the sample. Then a question level regression analysis is applied to identify the importance of several factors that theory suggests would affect the decision to become noncompliant.

Did public disclosure matter? A Chow test for structural break

As already noted, one of the characteristics of the BFC programme involves audits in which monitors enter the plants and record observations. These observations were the basis of BFC Synthesis Reports that were publically available on the internet. These reports named factories and linked them directly to working condition violations. The policy of posting these reports changed in November 2006, at which point Synthesis Reports stopped naming specific factories and only published aggregate compliance data.

This change in policy provides an opportunity to investigate the possible role that public disclosure has in plant manager behaviour. To investigate this change formally, first a Chow-type test is applied for a structural break in the decision to become noncompliant. The decision to become noncompliant is calculated at the compliance question level and is defined as a factory being observed as compliant and then subsequently observed to be noncompliant.

Figure 10.1 shows that two peak values emerge for this definition of the decision to become noncompliant. The peak values in both

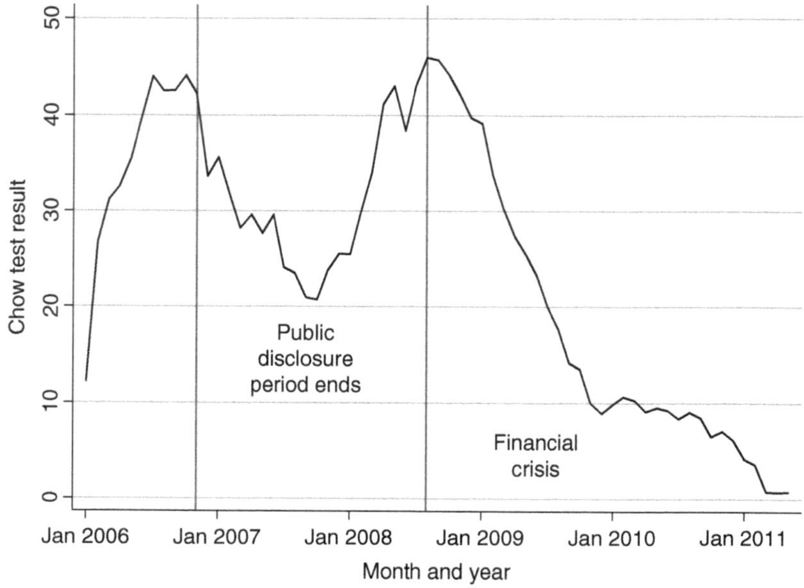

Figure 10.1 Chow-type test for structural break: general definition of regression

figures suggest that a structural break occurred around October 2006. Figure 10.1 also suggests that a second break occurred at the onset of the financial crisis. The fact that the data suggest a structural break at approximately the same time as the policy change supports the hypothesis that public disclosure affects the decision to become noncompliant. The fact that the break appears one month before the policy change would be consistent with some advance notice of the change occurring or that the break induced the policy change. Discussions with ILO/BFC management, however, suggest that the former is a much more likely explanation.

Regression analysis

Given the strong evidence supporting the possibility of a change in behaviour at the same time public reporting ceased, this section now turns to a more detailed regression analysis. In the regressions that follow a non-compliance variable is first created for each question so that the unit of observation is a factory question in each visit. Define $g_{qit} = 1$ if question q in plant i at time t changes from compliant to noncompliant and 0 otherwise. Using this as a dependent variable gives about 300,000 observations (questions * plants * visit).

The mean of the first reduced compliance dummy variable is 0.035. This 3.5 per cent of the question-plant-period observations can be decomposed to reveal falling rates of reduced compliance as the visit number increases. Between the first and second period the falling compliance rate (which is across all questions and factories) is 7.3 per cent, but that rate falls to 3.3 per cent between the fourth and fifth visit. Part of this may be explained by a selection bias if the least successful firms are most likely to reduce compliance and are more likely to drop out of the sample. These means are much smaller than the aggregate figures presented in Tables 10.2 and 10.4, which, through aggregation, contain composition effects.

In order to investigate the impact of buyer reputation sensitivity on labour law compliance, data were collected on each buyer's commitment to CSR, whether the firm is an apparel retailer or mass merchandiser, and other measures of brand value as determined by consulting firms such as Inter-Brand's Best Global Brands Ranking and *Fortune*'s 'Most Admired Companies' scoring system. Based on this survey of information, buyers were first separated into apparel retailers and mass merchandisers. Apparel retailers are primarily in the business of selling apparel and may sell other related but non-apparel goods. Mass merchandisers refer to large chain stores that sell a wide range of products, with apparel being only one subgroup. These two groups of buyers differ principally in terms of product quality measures both in terms of the technical characteristics of the garment and defect rate.

Within these two groups, buyers are subsequently divided by reputation sensitivity. Reputation-sensitive firms are characterized by evidence of a policy on CSR in the form of a website or public report. Of buyers sourcing from Cambodia during the study period, firms fell into four broad categories. The first (Type 1) includes apparel retailers with significant evidence of CSR. Apparel retailers with little evidence of a policy relating to CSR fall in the second group (Type 2). Type 3 buyers are mass merchandisers with significant evidence of CSR. No buyers fell into the category of mass merchandiser without evidence of CSR. The last type (Type 4) consists of buyers that were not accessing BFC compliance reports.

The simplest specification results, broken down by buyer type and using just the factors identified in the factor analysis, are found in Table 10.5. The regressions in all columns represent OLS estimation of a linear probability model of the binary dependent variable for the decision to become noncompliant that has been described here. The R-squared statistic is not reported because it has limited applicability in the linear

Table 10.5 Regression main factor groups – question level linear probability model

Variables	(1) Means full sample	(2) Means buyer type=1	(3) Means buyer type=3
Communication and workplace systems	0.040***	0.037***	0.040***
	[0.001]	[0.002]	[0.002]
Occupational safety and health	0.044***	0.039***	0.041***
	[0.001]	[0.002]	[0.002]
Modern HR practices	0.036***	0.033***	0.034***
	[0.001]	[0.002]	[0.002]
Compensation	0.027***	0.022***	0.026***
	[0.001]	[0.001]	[0.001]
Unions	0.010***	0.008***	0.010***
	[0.001]	[0.001]	[0.001]
Core labour standards	0.007***	0.006***	0.009***
	[0.001]	[0.001]	[0.002]
Observations	459,589	128,605	141,991

Notes: Robust standard errors in brackets. *** p<0.01, ** p<0.05, * p<0.10.

probability model. Probit estimation generates very similar results. The constant term is suppressed to allow the main effects of the various groups to each be represented as conditional means. In other words, the results in Table 10.5 simply represent the conditional means of the rates at which factories become noncompliant calculated at the question level for each of the groups identified in the factor analysis. These numbers differ from those in Table 10.2 because those were the group averages (which are in a sense cumulative within the group) rather than the question level averages.

The first thing to notice about the results in Table 10.5 is that the rates of becoming noncompliant are very low (less than 5 per cent in all cases). The second main message is that the overall results are very similar across buyer types. Considerable cross-factor variation occurs across the six factors. There is very little movement towards non-compliance in 'core labour standards', which, as described in the summary statistics section, are *zero-tolerance* points of compliance for buyers. Thus, it is not surprising to see few factories becoming noncompliant on these points. By contrast, becoming noncompliant in areas that involve more complex factory organizational change is considerably more common. The probability that BFC enterprise advisers would observe non-compliance on a visit following visits where the factory was found to be in compliance in 'communication and workplace systems', 'occupational safety and health' and 'modern HR practices' is close to 0.04.

Control variables

The means in Table 10.5 are largely descriptive and are meant to illustrate the differences across the different compliance areas, but omit variables that might affect the decision to become noncompliant. The first control variables added are geographic fixed effects that control for the nation of factory ownership. They are added to the regression as a set of dummy variables (one variable for each country identified in the data). Previous studies have found a significant effect of foreign ownership on technology adoption and survival (Harris and Li 2010).[4] Although the authors are not aware of studies that have included this variable, these effects are included in each of the columns of Table 10.6.

Several other variables that theory suggests might play a role are also added. Previous studies have found that a relationship with a reputation-sensitive buyer increases the propensity to improve working conditions (Oka 2010a). The first column of Table 10.6 shows that this relationship has a statistically significant effect on the decision to become noncompliant as well: a relationship with a reputation-sensitive buyer deters the decision to become noncompliant. This effect is relatively small, however, as suggested by the point estimate.

Adjustment costs can affect the regression. Here adjustment costs are modelled by classifying each of the areas in the survey with a dummy variable equal to one if the change implied represents a 'physically irreversible' decision (such as the purchase of capital or some other physical equipment that would be costly to remove), and zero otherwise. The results in Table 10.6 suggest that the presence of a physically irreversible compliance point has a statistically significant effect of reducing the decision to become noncompliant. This effect is quite similar across buyer types as well, and is consistently negative. This result is consistent with the investment literature that shows that irreversibility matters (Anderson et al. 2010; Mason and Weeds 2010).

It is also considered whether or not the threat of public disclosure of non-compliance can also deter the decision to become noncompliant. In Table 10.6, the coefficients of the 'public disclosure' variable are all positive, suggesting that the decision to become noncompliant became more common during the public disclosure period. The effect is relatively large and is not consistent with Ang et al. (2011), who find that

Table 10.6 Regression main factors with controls

Variables	(1) Full sample	(2) Buyer type=1	(3) Buyer type=3
Communication and workplace systems	0.052***	0.039***	0.051***
	[0.003]	[0.005]	[0.004]
Occupational safety and health	0.059***	0.043***	0.055***
	[0.003]	[0.004]	[0.004]
Modern HR practices	0.046***	0.033***	0.043***
	[0.003]	[0.005]	[0.004]
Compensation	0.038***	0.022***	0.035***
	[0.003]	[0.004]	[0.004]
Unions	0.019***	0.008*	0.018***
	[0.003]	[0.004]	[0.004]
Core labour standards	0.017***	0.005	0.017***
	[0.003]	[0.004]	[0.004]
Reputation-sensitive buyer	–0.006***	n.a.	n.a.
	[0.001]	n.a.	n.a.
Physically irreversible compliance point	–0.015***	–0.014***	–0.017***
	[0.001]	[0.002]	[0.002]
Union active in labour rights	–0.004*	–0.001	–0.007*
	[0.002]	[0.004]	[0.004]
Small unions possibly controlled by management	–0.005*	–0.001	–0.009*
	[0.003]	[0.006]	[0.005]
Unions known to be politically affiliated	–0.013***	–0.006	–0.026***
	[0.004]	[0.004]	[0.005]
Large unions known to serve management	0.001	0.004	–0.001
	[0.003]	[0.006]	[0.005]
Public disclosure period	0.030***	0.020***	0.033***
	[0.004]	[0.007]	[0.007]
Crisis period	–0.007***	–0.003	–0.010**
	[0.002]	[0.004]	[0.004]
Recovery period	–0.006**	–0.003	–0.004
	[0.002]	[0.004]	[0.005]
Observations	445,817	122,322	139,324

Notes: Robust standard errors in brackets. *** $p<0.01$, ** $p<0.05$, * $p<0.10$. Buyer types are explained in the text.

compliance generally was higher during the public disclosure period. It is important to note, however, that compliance rates do not fall to initial (first visit) levels for any factory. Relative to overall compliance and to previous improvements in compliance, the rates of becoming noncompliant are low. This result is interpreted here to suggest that, while some firms may have been motivated to comply in some areas due to public exposure, the majority of factories had additional motivation for most compliance points.

Four union variables are also included. Unions are particularly interesting in Cambodia because they may either represent workers, and, therefore, support improvements in working conditions (holding all else constant), or they may be aligned with either factory managers or political parties that either support or do not support such improvements (see Arnold, Chapter 9 in this volume for an in-depth analysis). The estimates in Table 10.6 show very few statistically significant union coefficients. This may be due to large but heterogeneous union effects (therefore, giving rise to large standard errors) or very small effects of unions (small estimated coefficients). The estimated coefficients in Table 10.6 are quite small, suggesting that unions have very little effect on the decision to become noncompliant. Unions active in labour rights are the only ones with statistically significant coefficients, and these are both positive (in columns (1) and (3) of Table 10.6). The positive coefficients suggest that the presence of these unions increases the likelihood of the decision to become noncompliant, which is a result that merits additional research.

Finally, controls for the financial crisis (June 2008–December 2009) and recovery period (beginning in January 2010 and extending to the end of the sample period) are also included. During the financial crisis demand from developed countries dropped considerably (Baldwin 2009), causing Cambodian exports to drop sharply, which placed increased pressure on factories. Factory closures increased significantly. The crisis, therefore, offers an opportunity to consider the effects of an exogenous adverse demand shock. If sustaining improvements is costly to factories, then it is possible that the decision to become noncompliant would become more common during the crisis. A negative estimate could be consistent with positive effects of working conditions such that non-compliance is less attractive during the crisis.

The crisis coefficient estimates in Table 10.6 are negative, suggesting that the decision to become less compliant became less common during the crisis. These results seem inconsistent with the hypothesis that reversing improvements in working conditions helped factories during adverse economic times because they were burdensome. Given that there were significant productivity improvements in Cambodian apparel factories that occurred along with improvements in working conditions (Asuyama et al. 2010), it seems possible that the two improvements might be related.

One concern about the results in Table 10.6 is that unobserved firm-specific characteristics could be driving these results. For example, it is possible that more able managers are more likely to be able to make improvements in working conditions that increase the productivity of the factory and, therefore, they are less likely to retrogress. One advantage of these data is that factories are followed over time, which enables the inclusion of factory-specific fixed effects.

Table 10.7 contains the results that include factory-specific fixed effects. This set of dummy variables (one for each factory) controls for

Table 10.7 Factory-level fixed effects

Variables	(1) Full sample	(2) Buyer type=1	(3) Buyer type=3
Communication and workplace systems	0.060***	0.036***	0.056***
	[0.005]	[0.005]	[0.003]
Occupational safety and health	0.067***	0.040***	0.060***
	[0.005]	[0.004]	[0.003]
Modern HR practices	0.054***	0.030***	0.049***
	[0.005]	[0.005]	[0.003]
Compensation	0.046***	0.019***	0.040***
	[0.005]	[0.004]	[0.003]
Unions	0.028***	0.004	0.023***
	[0.005]	[0.005]	[0.003]
Core labour standards	0.025***	0.002	0.022***
	[0.005]	[0.005]	[0.003]
Physically irreversible compliance point	–0.015***	–0.014***	–0.017***
	[0.001]	[0.002]	[0.002]
Union active in labour rights	–0.003	–0.004	–0.005
	[0.003]	[0.005]	[0.005]
Small unions possibly controlled by management	–0.004	–0.004	–0.002
	[0.003]	[0.006]	[0.005]
Unions known to be politically affiliated	–0.018***	0.002	–0.019***
	[0.004]	[0.004]	[0.004]
Large unions known to serve management	0.003	0.003	0.002
	[0.004]	[0.006]	[0.005]
Public disclosure period	0.032***	0.026***	0.036***
	[0.004]	[0.008]	[0.006]
Crisis period	–0.005**	–0.003	–0.006
	[0.003]	[0.005]	[0.004]
Recovery period	–0.003	–0.003	0.000
	[0.003]	[0.004]	[0.005]
Observations	445,817	122,322	139,324

Notes: Robust standard errors in brackets. *** $p<0.01$, ** $p<0.05$, * $p<0.10$.

any factory characteristic that is constant over time (such as a manager or factory owner that remains in his or her position for the entire sample). The results with these controls are very similar to those in Table 10.6, suggesting that Table 10.6 results are not driven by unobserved factory-specific characteristics. In particular, physically irreversible compliance points and the crisis reduce the decision to become noncompliant. The union variables remain largely insignificant but follow similar patterns as those suggested in Table 10.6.

Conclusions

Using data from the BFC programme, this chapter estimates the rates of, and factors that affect, factory level decisions to become noncompliant on a very wide range of individual compliance points. One of the main conclusions of this study is that rates of becoming noncompliant are extremely low among Cambodian garment factories; a result that seems inconsistent with the idea that improvements in working conditions are especially burdensome for factories. The rates of becoming noncompliant are lowest in the areas of 'core labour standards' and 'unions'. 'Compensation' follows these areas with much lower rates of becoming noncompliant than other areas such as 'communication and workplace systems'. The low rates of becoming noncompliant in this area are consistent with the 'efficiency wage' idea that offering consistent compensation to workers may help improve productivity and increase factory performance.

The results identify several factors that may affect a factory's decision to become noncompliant with a very large number of working conditions. An important determinant of a factory's decision to become noncompliant is adjustment costs. Compliance areas that require an investment, or that are costly to reverse, are less likely to be reversed. While this may seem like a trivial point, the implication of this is that there are points that might be beneficial to the factory (that would bring the benefits that compliance brings), but not implemented by the factory because the certainty of the cost and the uncertainty of the benefit combine to make managers hesitant to implement such changes. When given additional incentive to make these investments, such as through the Better Factories programme, factories may 'take the leap' and then later discover that the changes were indeed beneficial. The large and growing investment literature focusing on the dynamic effects of irreversibility suggests that the strategic implications of the effects of irreversibility merit attention in future research.

In terms of policy recommendations, this chapter has important implications. Although improving human resource policies may have positive benefits for the factory, they may not be implemented if factories are certain about the one-time costs of implementing such improvements but are uncertain about the benefits. If this is the case, and the results in this chapter are consistent with this hypothesis, then the policy implication is that subsidizing one-time investments and sharing information about the specific reforms that factories have tried and have proven to be beneficial would help other factories overcome these obstacles and lead to more beneficial improvements in working conditions.

Notes

1. Some regulations seem to push factories so far from their optimal practices that they increase the probability of closure (Biørn et al. 1997). There are very few papers that examine the link between factory survival and improvements in working conditions; Harrison and Scorse (2010) is one notable exception.
2. Prominent examples include Polaski (2004 and 2006), Berik and van der Meulen Rogers (2010) and Oka (2010a). More information can be found at http://www.betterfactories.org/ (accessed 13 August 2013).
3. The principle-components factor method is a common alternative, but this method assumes that the commonalities are equal to one. The average of our uniqueness estimates is just over 0.65, and the principle-components method is most appropriate for uniqueness values close to zero. In this case, therefore, the principle components analysis is probably not appropriate.
4. Brown et al. (2011b) examine improvements in working conditions and factory survival in Cambodian apparel factories.

11
Towards Better Work in Central America: Nicaragua and the CAFTA Context

Jennifer Bair and Gary Gereffi

Introduction

Like many other low-income industrializing economies, Nicaragua's export sector relies heavily on apparel manufacturing for revenues and employment. However, at least among the major garment-producing countries in the Western hemisphere, Nicaragua is unique: unlike the majority of its neighbours in Latin America, its apparel industry has expanded since 2005. In this sense, Nicaragua's experience counters a general trend in the global garment industry towards greater concentration in Asia, and particularly China, the world's largest clothing exporter.

Nicaragua is also the only country in Central America, and the second in the Americas (following Haiti), to participate in the Better Work programme. Here, too, the Nicaraguan case is exceptional because Nicaraguan garment factories are not associated with systematic abuses of workers' rights. The government's record of labour law enforcement, and the industry's compliance with those laws, at least in recent years, has been relatively positive. Consequently, the Nicaraguan government viewed participation in the Better Work programme mainly as an opportunity to publicize what it perceived as the country's existing strengths as a 'high road' exporter.

Nicaragua's participation in Better Work comes at a critical time in the evolution of its garment sector. Growth in apparel exports has been fuelled by the country's participation in the Dominican Republic-Central American Free Trade Agreement (CAFTA-DR) with the United States, which was signed by the governments of Nicaragua and five other countries in the region in 2004. While duty-free access to the US market has provided the CAFTA signatories an advantage vis-à-vis other apparel exporters, for most of the CAFTA countries this advantage has not been sufficient to offset the greater competitiveness of Asian suppliers.

A decisive factor permitting the expansion of Nicaragua's apparel exports to the United States is the fact that Nicaragua was granted preferential treatment under CAFTA in recognition of its status as the least developed among the participating countries. Specifically, Nicaragua was granted a limited volume of tariff preference levels (TPLs), which enable garments assembled in Nicaragua from materials that do not meet the yarn-forward rules of origin established under the CAFTA (e.g. fabrics originating in Asia instead of the Americas) to access the US market duty-free. Although it is unclear precisely how much of Nicaragua's export dynamism is attributable to TPLs, they have undoubtedly helped to fuel the sector's growth. However, the TPL preference granted to Nicaragua under CAFTA is a temporary one, and it is set to expire in 2014.

The current climate of uncertainty in Nicaragua created by the scheduled expiration of the TPLs casts into stark relief the trade-dependent nature of its development strategy. The competitiveness of its chief manufactured export – clothing – is contingent on trade preferences that are outside the control of either the Nicaraguan government or the private sector. This situation raises important questions about the relationship between economic upgrading (the process of increasing competitiveness in foreign markets) and social upgrading (the process of generating more and better jobs) in global value chains (see Bernhardt, Chapter 2 in this volume). At present, Nicaragua's ability to leverage participation in the apparel value chain into economic and social upgrading appears contingent on the regulatory environment created by the CAFTA regime. If the TPLs are not extended and the industry is unable to adjust to the loss of these preferences, economic upgrading may be stalled or even reversed. This, in turn, is likely to affect negatively the broader agenda of social upgrading since it is difficult to improve wages, working conditions and employment security in an export sector dominated by a shrinking industry.

How might Nicaragua benefit from its engagement with the Better Work programme, and conversely can Better Work use its experience in Nicaragua to find new ways to improve productivity and labour outcomes in the apparel global value chain? Can Better Work strengthen the tenuous connection between economic and social upgrading by creating more stable export growth prospects with foreign buyers? Given that Better Work is a partnership between the International Labour Organization (ILO) and the World Bank's private sector financing arm, the International Finance Corporation (IFC), how can this unique structure advance Better Work's objectives in developing countries?

This chapter argues that Nicaragua presents a challenge as well as an opportunity for Better Work to move beyond a model of monitoring factory level compliance in favour of an approach that identifies the root causes of non-compliance at the firm level. In order to meet this challenge and seize the opportunity that Nicaragua provides, Better Work needs to develop an innovative and dedicated approach to stakeholder engagement with the goal of enlisting participating brands as genuine partners in the search for decent work. If Better Work Nicaragua instead opts for a narrower agenda (e.g. monitoring factories for compliance with international standards and domestic labour law, but leaving other issues, like buyer–supplier relations, unaddressed), then the value of Better Work to the prevailing (and generally inadequate) system of social compliance audits is limited.

Mapping Central America in the global value chain

The MFA phase-out and the removal of quotas on textile products on 1 January 2005 marked the end of over 30 years of restricted access for developing country exporters to the markets of the European Union (EU) and North America. Retailers and other buyers became free to source apparel in any amount from any country, subject only to a system of tariffs and a narrow set of transitional safeguards that were set to expire at the end of 2008. This caused a tremendous shift in the global geography of apparel production and trade and a restructuring of firm strategies, as companies sought to realign their production and sourcing networks to accommodate new economic and political realities (Tewari 2006).

In tandem with the liberalization of global garment trade, regional trade agreements have also played a major role in strengthening ties between the United States, the largest apparel market in the world, and its main trading partners. The North American Free Trade Agreement (NAFTA), which was signed in 1994, and CAFTA, which was signed a decade later, aimed to improve the competitiveness of the US textile industry and apparel exporters from Mexico and the Caribbean Basin in the face of rapid growth of low-cost apparel exports coming primarily from Asia (see Frederick and Gereffi 2011; Gereffi, Spener and Bair 2002).[1]

Table 11.1, which shows the growth of US apparel imports from 1990 until 2011, tracks the rise and fall of various apparel suppliers. Total

Table 11.1 US apparel imports: regional and Asian suppliers, 1990–2011

Partner	Value (in US$ millions)							% of total value			
	1990	1995	2000	2005	2009	2010	2011	'90	'00	'10	'11
World	21,937	34,649	57,232	68,713	63,105	71,398	77,659				
China	2,739	3,518	4,499	15,143	23,503	27,975	29,392	12	8	39	38
CAFTA-DR	1,434	4,745	8,973	9,104	6,145	7,016	7,853	7	16	10	10
Viet Nam	0	17	47	2,725	5,068	5,877	6,644	0	0	8	9
Bangladesh	429	1,067	2,116	2,372	3,410	3,930	4,510	2	4	6	6
Mexico	508	2,566	8,413	6,078	3,391	3,541	3,804	2	15	5	5
Cambodia	0.1	0.5	808	1,713	1,871	2,222	2,592	0	1	3	3
Total								**23**	**43**	**71**	**71**

Note: % represents a country or region's market share of the total value of US imports of apparel in a given year.
Source: US Department of Commerce, Office of Textiles and Apparel (OTEXA): imports by country by MFA category: category 1: all apparel.

US apparel imports more than tripled between 1990 and 2005 from US$21.9 to US$68.7 billion. Although imports fell between 2005 and 2009, reflecting the impact of the deep global economic recession in 2008–2009, by 2011 they had rebounded to US$77.7 billion. China was the leading exporter at the beginning of this period (US$2.74 billion), but the CAFTA countries collectively were in second place, with 52 per cent of China's apparel export total in 1990. However, China's US apparel exports accelerated much faster than its rivals did after 2005, with its total share of the US market in 2011 nearly quadrupling the import market share claimed by the CAFTA countries. China exceeded Mexico's US market share in the same year by a factor of eight.

From a regional perspective, therefore, Mexico and Central America both experienced sharp declines in their share of US apparel imports between 2000 and 2011: Mexico's share fell from 15 to 5 per cent, while the share of the CAFTA-DR countries, taken as a group, decreased from 16 to 10 per cent. During this same period, China enjoyed dramatic growth in its share of US apparel imports, which rose from 8 per cent in 2000 to 22 per cent in 2005 and 38 per cent in 2011. Viet Nam also burst onto the scene during the past decade, going from virtually no apparel exports to the United States in 2000 to a US import share of 4 per cent in 2005 and 9 per cent in 2011. While CAFTA-DR and Mexico have been losing US import market share since 2000, China, Viet Nam, Bangladesh and Cambodia were all gaining ground.

One of the key questions for Mexico and Central American countries is the degree to which NAFTA and CAFTA would maintain or even strengthen the position of regional exporters in a liberalized global garment trade. While Mexico's US import market share has fallen dramatically in the past decade, the decline has been both more gradual and uneven within the CAFTA region. CAFTA has helped maintain the position of Central American and Caribbean exporters among leading suppliers of apparel to the United States, although the value of the region's exports to the United States fell from US$9.1 billion in 2005 to US$7.9 billion in 2011. Within CAFTA, the Dominican Republic and Costa Rica witnessed significant declines in their exports to the United States, but these declines have been offset by growth in shipments from Honduras, El Salvador and, most recently, Nicaragua.

As Table 11.2 shows, in both 2005 and 2011 Honduras ranked first among CAFTA exporters to the United States. El Salvador and Nicaragua now rank second and third, with Nicaragua edging slightly ahead of Guatamala in the most recent year. Nicaragua's exports to the United States nearly doubled in value between 2005 and 2011; all other

Table 11.2 US apparel imports from CAFTA countries, 1995–2011

Country	Value (in US$ millions)					% of CAFTA-DR Value				
	1995	2000	2005	2010	2011	'95	'00	'05	'10	'11
CAFTA-DR	4,745	8,973	9,104	7,016	7,853					
Honduras	919	2,323	2,622	2,414	2,615	19	26	29	34	33
El Salvador	582	1,583	1,619	1,638	1,738	12	18	18	23	22
Nicaragua	74	336	716	1,017	1,357	2	4	8	14	17
Guatemala	682	1,487	1,816	1,152	1,321	14	17	20	16	17
Dominican Republic	1,731	2,425	1,849	626	654	36	27	20	9	8
Costa Rica	757	819	482	168	167	16	9	5	2	2

Source: U.S. Department of Commerce, Office of Textiles and Apparel (OTEXA): MFA category 1: all apparel imports.

countries in the CAFTA region except El Salvador declined during this period. While in 2005 Nicaragua accounted for only 8 per cent of the region's apparel exports to the United States, by 2011 that percentage had increased to 17 per cent.

Despite this strong performance, CAFTA regulations have created an environment of uncertainty in Nicaragua due to the pending expiration of TPLs. In recognition of the lower cost, greater availability and sometimes better quality of Asian fabrics, CAFTA allows Nicaragua to receive preferential access to the US market for a certain quantity of apparel (up to 100 million square metre equivalents) sewn in Nicaragua from materials that do not meet CAFTA's rules of origin. The CAFTA granted Nicaragua TPLs for a ten-year period, and they are due to expire in 2014. Nicaragua was the only CAFTA country to receive TPLs, which have been crucial, given the absence of domestic textile production in the country and the limited availability of cost-competitive fabrics produced in the Americas.

In 2009, 82 per cent of Nicaragua's exports to the United States entered the country duty-free under a variety of different special trade regimes. Over one-third of exports (35 per cent) entered under the regional rules of origin established by CAFTA, while 47 per cent of exports were imported under the TPLs granted to non-originating exports. As these figures suggest, Nicaraguan manufacturers are heavily reliant on TPLs. The CAFTA's yarn-forward rule of origin, combined with the looming expiration of the TPLs, is generating alarm on the part of local stakeholders that US buyers will shift orders to other suppliers, especially in Asia, when the TPL benefit expires. As explained in this chapter, these fears are exacerbated by the fact that a sizable percentage of the companies that are active in Nicaragua are subsidiaries of foreign companies with a global presence, including factories in Asia to which they could redirect orders should they decide to cease production in Central America.

Results

This chapter is based on data gathered during field research that was conducted by the authors during fall 2010 and summer 2011.[2] A total of 52 interviews were carried out for this project, of which 29 were with companies; 13 with government agencies, including the National Free Zone Commission (CNZF), the Ministry of Labour and PRONicaragua, the government's foreign investment promotion arm; and the remaining ten with other stakeholders, including Better Work staff, the apparel industry association and other employer organizations, trade union officials and labour rights experts.[3]

The CNZF directory was used to identify a representative sample of companies in terms of national origin, product mix and size, although larger firms were oversampled. With the exception of two (a converter of woven fabrics and a new agro-enterprise for cotton), these companies were engaged in the manufacture of either woven or knit apparel. In numerical terms, they represented approximately 40 per cent of the total number of garment firms operating in Nicaragua under the CNZF regime, but 79 per cent of total apparel manufacturing jobs. In this section, the findings from the fieldwork are summarized, addressing first the general themes of economic and social upgrading that are central to this volume before describing a distinction within the apparel value chain that was identified between 'Americas-based' and 'Asia-based' segments.

Economic upgrading

Many studies of the global value chain for apparel differentiate between 'full-package' production, in which suppliers coordinate all aspects of production and purchase the fabric used in the client's orders, and assembly subcontracting, in which the client finances the raw materials and pays the supplier simply for cutting the fabric and assembling the garment. Furthermore, several scholars describe the transition from assembly subcontracting (also known as CMT for 'cut, make and trim') to full-package manufacturing as a form of economic upgrading (Bair and Gereffi 2001; Gereffi 1999). Based on empirical evidence from cross-regional comparisons of exporters in East Asia and Latin America, as well as case studies of particular exporting clusters in several countries, they claimed that full-package manufacturing provides suppliers with increased competitiveness and larger margins than traditional assembly subcontracting. Because this production model may give local suppliers more control over links in the value chain beyond assembly – for example, the sourcing of inputs and post-assembly processes such as laundering and finishing – they further reasoned that it is more likely to encourage the development of backward and forward linkages to other kinds of activities; thus, stimulating investment and employment across different segments of the chain.

The Nicaraguan industry hosts a mix of full-package and assembly production. The salient distinction in this regard is between knit and woven apparel.[4] All of the companies in the knit sample are either full-package producers, meaning that they finance the purchase of fabric used in clients' orders, or they manufacture their own brands of apparel in addition to doing some subcontracting of private label (store brands) for retailers. Among manufacturers of woven garments, a greater mix of assembly subcontracting and full-package production was found. The greater prevalence of the full-package model among knitwear companies reflects the fact that these manufacturers are much more likely to be vertically integrated producers that convert the fabric they produce into garments. For example, five of the knitwear companies we interviewed are using a regional production model that involves Nicaraguan sewing factories and Honduran knitting mills.[5] Several others have knitting mills in Asia that produce the fabrics assembled in Nicaragua.

Contrary to the stylized upgrading trajectory that assumes a move from assembly subcontracting to full-package production (Gereffi 1999), two of the companies interviewed that manufacture woven garments have moved in the opposite direction. Full-package production became too expensive for these companies to sustain, given the rising costs of fabric and the lack of accessible, affordable credit to finance these textile purchases. Several of the firms in the sample reported that they were unable to access credit from local banks, and because their assets (namely, their factories) were located in Nicaragua they were unable to secure financing from foreign banks. The credit crunch these manufacturers face is exacerbated by client purchasing practices, since apparel suppliers may have to wait a month or more for payment from the foreign brands whose orders they fill. Consequently, significant working capital (in excess of US$1 million) is required to finance the textile purchase for even a relatively modest full-package order.

Beyond the particular issue of financing, several companies expressed scepticism about the merits of full-package production versus assembly subcontracting. They pointed out that full-package manufacturers incur much greater risks, since they have no alternative but to absorb the cost of goods returned or rejected by the client – for example, for poor quality. Although full-package production may be more profitable than traditional assembly subcontracting, not all manufacturers are convinced that the margins are sufficiently high to offset this greater risk. Furthermore, the firms interviewed reported that they do not necessarily have greater control over the production process when filling a full-package order due to the detailed specifications client firms provide.

For example, when placing an order buyers may specify not just the particular fabric they want the manufacturer to use, but even the particular vendor from which the textiles should be purchased. This level of buyer oversight limits the degree to which firms can purchase from local producers, even when such inputs are available. Client decisions about fabric sourcing are particularly critical in the Nicaraguan context, given the limited availability of TPL for non-originating fabrics. In short, these findings support the view that while full-package production may be a necessary condition of economic upgrading in the post-MFA apparel industry, it is not a sufficient condition of sustained competitiveness nor of social upgrading (Bair and Werner 2011; Plankey-Videla 2012; Schrank 2004).

Given some of the disadvantages of full-package production, why is it still the dominant production model in Nicaragua? Simply put, apparel manufacturers report that the ability to finance and coordinate full-package orders is a prerequisite for doing business with many clients. Much of the production carried out in Nicaragua is for the private label lines of mass retailers, such as Walmart, Target and Kohl's. These companies want to place orders with suppliers that can deliver a 'full-package'. Accordingly, buyers expect their suppliers to do more than just assembly or CMT, and the firms interviewed in Nicaragua are no exception. The majority offer additional services, most typically the laundering that is a standard part of the production process for jeans and twill pants, or, in the case of knitwear manufacturers, screen-printing and embroidery. Several companies also provide more sophisticated pre-production activities, including pattern marking and grading, and some product development. These findings indicate economic upgrading, as Nicaragua's value chain lengthens to include links beyond sewing.

Nevertheless, for a number of firms the additional value generated by expanded capabilities was offset by increased cost pressure due to the stagnant, or in some cases even declining, prices paid by clients. Two companies put the decline in prices over the last several years at 10–20 per cent. Many were unable to pass on rising production costs to foreign buyers. At the time of the authors' fieldwork, firms reported having been negatively affected by a dramatic spike in the price of cotton (and, thus, cotton textiles), which began during the second half of 2010 and continued into 2011 before declining. Local factors also contributed to increased cost pressures. Several firms noted the rise in labour costs caused by a series of government-mandated minimum wage increases that occurred under the current presidential administration, prior to the negotiation of the first Tripartite Agreement in 2009 (discussed below).

Yet, in spite of these challenges, Nicaragua's share of the global apparel market has been growing. The trajectories of several of the firms interviewed reflect the export performance described here. Numerous firms indicated that that their plants were operating at 100 per cent production capacity at the time of the fieldwork. One company had actually grown substantially over the preceding year from 300 employees to 1,400 employees, and another three were planning expansions that would increase production volumes by between 20 and 50 per cent. While Nicaragua fulfils one of the two criteria for economic upgrading described by Bernhardt in Chapter 2 of this volume (increased market share), it fails to meet the other: the unit value of its exports declined between 1990 and 2009. For this reason, Nicaragua is classified by Bernhardt as an intermediate case of upgrading, having achieved an increase in export volumes without a corresponding increase in the value of its exports.

Social upgrading

Assessing the degree to which social upgrading is occurring in Nicaragua's apparel industry is a challenging task, particularly in the absence of macro-comparative data on wages and working conditions that could be used to benchmark Nicaragua against other countries over time. In purely quantitative terms, the picture is clear: the number of jobs has increased substantially. When moving from quantitative to qualitative metrics, however, the situation becomes more complex.

The authors found that labour issues were a far less acute concern for firms than trade-related developments, principally the scheduled expiration of the TPLs in 2014. Although the social compliance of garment exporters has been a matter of public debate, the companies interviewed expressed relatively little concern about the costs and pressures of complying with either the client's standards or Nicaraguan labour law. Virtually all reported being subject to client codes of conduct, but less than half of the companies in the sample were WRAP-certified (Worldwide Responsible Accredited Production is an industry-organized certification system). One was an FLA (Fair Labor Association) participating supplier, and another was going through FLA certification at the time of the interview in summer 2011.

Manufacturers expressed frustration with the redundancies inherent in the current social compliance system: because factories generally produce for multiple buyers, each with its own code of conduct and procedures for monitoring performance, many factories undergo audits multiple times throughout the year. Several firms noted that the principal reason they would consider participating in Better Work was the decision of some brands and retailers to accept the results of Better Work audits in place of their own.

A striking finding from the firm-level interviews with regard to labour was the relatively high rate of turnover. For knit firms, turnover among production workers was especially volatile. On the low side, one US-owned company in the sample that uses a modular production system to produce a variety of complex intimate garments in relatively small volumes reported turnover of less than 12 per cent per year. In contrast, other manufacturers had turnover rates of between 120 and 180 per cent annually, and one acknowledged an annual turnover rate of 300 per cent. No clear pattern emerged with regard to a correlation between turnover rates and reported wages, the presence or absence of a union or the location of the factory (urban versus rural area). Although the small sample size makes it difficult to generalize about the factors that might explain either very low or very high turnover rates, it appeared that companies offering more generous fringe benefits, such as transportation to and from the factory and subsidized lunch in an on-site cafeteria, enjoyed lower turnover.

Unsurprisingly, there was a correlation between turnover rates and productivity: companies having turnover rates above 60 per cent a year had lower productivity than their counterparts with lower turnover. One company located in a relatively rural area several hours from the capital city, Managua, reported that managing high rates of turnover was particularly challenging in the context of a small local labour market. Management at this factory had recently begun conducting exit interviews with workers, and found that many were leaving to pursue educational opportunities in Managua. Others migrated to Costa Rica in search of seasonal agricultural work. Particularly in urban areas, managers regarded substantial rates of turnover as a typical and perhaps inevitable feature of the export sector. Those in Managua, where several large trade zones hosting multiple factories are located, have found it relatively easy to replace departing workers with new ones, many of whom already have experience in the industry.

Perhaps for this reason, some companies offer little in the way of formal training, and two of the firms manufacturing knit shirts reported that they provide none at all. Training varied across other factories from an average of two to ten weeks, with one outlier reporting that the training period for more complicated operations could be as long as 40 weeks. Nicaragua's national training institute – the National Institute of Technology (INATEC) – provides training and certification services for the workforce. Companies, including those in the free trade zone sector, pay fees equivalent to 2 per cent of their payroll to INATEC, which are intended to cover access for their employees to INATEC's technical education and training programmes. However, few of the companies interviewed were using INATEC's services at the time of this research. The consensus is that there is little connection between the kinds of skills that are in demand among apparel and textile firms, particularly of a technical or managerial nature, and the courses that INATEC offers. For example, several managers claimed they were constantly looking for supervisors and workers with technical skills, but these are not areas in which INATEC's programmes are strong.

Not all companies shared this assessment, however. Numerous firms reported that they had used INATEC's services in the past and found them satisfactory. The manager of a firm producing uniforms reported that his company had used INATEC for training new line supervisors. He further noted the significant pay increase associated with a promotion from sewing machine operator to line supervisor (from 4,500 cordobas to 10,000 cordobas per month). Although this company had three individuals on its managerial staff who began working in the factory as sewing machine operators, internal mobility was more the exception than the norm among local companies. In general, high turnover, paired with relatively modest levels of training, suggest that human capital formation and skills development among Nicaragua's garment workers are modest. This is an area of social upgrading with ample room for improvement.

The firms interviewed expressed concern about increasing labour costs, particularly in light of the minimum wage increases implemented by the government prior to 2009. However, they expected that increases in labour costs would be less dramatic in the immediate future, thanks to the so-called Tripartite Agreement. This Agreement, which was negotiated between the Nicaraguan government and representatives of the private sector and organized labour, establishes a set schedule of minimum wage increases through 2013.[6] In exchange for locking in minimum wage increases, the Agreement also calls for the government and private sector to establish commissaries that provide workers with basic commodities, such as cooking oil, beans and rice, at below market prices. Companies were generally positive about the Tripartite Agreement, seeing it as a proactive effort on the part of the government to create a more predictable environment for local firms with regard to production costs.[7]

However, it is unclear to what degree workers benefit from the Agreement, which essentially removes wage determination from the market and potentially locks in increases lower than would be generated by a tighter labour market, or by continuing political and social pressures. Furthermore, when the first Agreement was signed in 2009, stakeholders feared that the industry was entering a protracted downturn caused by slumping demand from foreign buyers impacted by the US recession. However, the dip in exports that motivated this concern turned out to be temporary, and the current Agreement, which extends the schedule of pre-established minimum wage increases from two to three years, was concluded during a period of robust growth.

The Tripartite Agreement is not without its critics, who claim that the social programmes promised to workers in exchange for concessions on wage increases (e.g. an affordable housing programme) have not developed as promised (Rogers 2012). Yet, while there are complaints about the implementation of specific clauses, there is also widespread support for the principle of social dialogue and the tripartite structure in place to facilitate it. With the current Agreement set to expire in 2013, government officials in 2012 were optimistic that a new agreement would follow. It was not yet clear what the duration of it would be, however, since some parties are arguing for a five-year agreement, while labour leaders who are wary of locking in negotiated, automatic wage increases for a longer period of time, want another three-year deal.

In the authors' interviews with firms, labour representatives and a variety of apparel industry stakeholders, there was broad consensus that the industrial relations environment in Nicaragua has improved markedly in recent years. The previous administration was perceived as lax in its enforcement of labour law and tolerant of employer practices, such as union busting and blacklisting, which had a chilling effect on workers' associational rights. However, there are lingering concerns about the ability of workers to exercise their rights to freedom of association and collective bargaining. Although it appears that violations of these rights, where they occur, are primarily a firm-level phenomenon and do not necessarily indicate a pervasive or industry-wide anti-union culture, compliance with these rights is now being assessed by Better Work Nicaragua and, thus, more data regarding the pervasiveness and severity of this problem may be available in the future.

Americas-based versus Asia-based segments in Nicaragua's apparel value chain

One of the most salient findings that emerged from this fieldwork is that Nicaragua participates in two different segments of the apparel value chain – one based in the Americas and the other in Asia. Unlike many other apparel exporting countries, such as the Dominican Republic, India, Mexico and Sri Lanka, domestic capital plays a negligible role in Nicaragua's export sector. Of the 27 firms interviewed, only one is locally owned. The majority are subsidiaries of parent companies located in North or Central America, including in Canada, El Salvador, Honduras, Mexico and Trinidad. In addition to this set of American-based firms, Nicaragua is also home to companies based in Asia. Korean investors are the most common among this group, although manufacturers with parent firms in Hong Kong Special Administrative Region and Taiwan (China) were also interviewed. This distinction between American and Asian ownership was significant in four respects.

First, the companies from Asia are concentrated in the knit segment of the apparel industry and manufacture mostly basic, low value-added garments, such as t-shirts, with the exception of two manufacturers of woven pants based in Taiwan (China). Korean manufacturers have a sizable presence in the knitwear segment, owning five of the 14 knitwear firms in Nicaragua. In contrast, companies from the United States dominate the production of woven apparel (eight of 13 firms).

Second, the Asian-owned companies are larger. The average number of employees reported by subsidiaries of firms based in Asia (3,638 workers) was more than double the size of subsidiaries with parent companies based in the Americas (1,560 workers). There is significant variation within each group, however, and the major dimension in terms of size appears to be type of apparel (knits versus woven) rather than ownership. The three largest employers in the free trade zone sector are all knitwear producers, and together they represent almost one-quarter of total apparel employment in Nicaragua's free trade zones.

Third, the Asian companies in the sample tended to be more global than their American counterparts. While many of the companies based

Table 11.3 Key indicators of firms in Nicaragua, 2011

Firm type[a]	Year est.[b]	Ownership[c]	Product type[d]	Fabric[e]	Product/ week[f]	Emp.[g]
K1	n.a.	United States	OBM, M	55–60% United States, rest from Asia, Mexico, Central America	240,000	1,300
K2	2001	United States	FP	United States CAFTA countries	200–250,000	900–1,300
K3	2006	United States	97% FP, 3% CMT	very little US fabric, Miliken, El Salvador	100,000	1,400
K4	2004	Republic of Korea	FP		600,000	5,200
K5		Canada	OBM	Honduras (US yarn)	3 million	5,500
K6	1994	Republic of Korea	FP	60% Taiwan (China) and China, 40% Honduras, minimal Guatemala	n.a.	5,600
K7	2002	Hong Kong (China)	FP	100% China (own textile mill)	75,000	700
K8	2005	United States	FP	Honduras (US, Pakistani yarn)	750,000	1,250
K9	2010	Republic of Korea	FP	70% Asia (mostly China); 30% Honduras	125,000	2,100
K10	1999	Republic of Korea	FP	Republic of Korea and China	475,000	2,777
K11	2007	Honduras	FP	90% Honduras (own textile mill, some Pakistani yarn)	120,000	680
K12	2008	El Salvador	FP	China, El Salvador, Guatemala	15–20,000	1,075
K13	2008	United States	FP	China, US, Guatemala, Honduras	5,000	330
K14	2005	Republic of Korea	FP	Republic of Korea, China	125,000	1,250
W15	2009	Nicaragua	90% CMT; 10% FP	50% Asia 50% US	125,000	3,900
W16	n.a.	United States	CMT	US, Mexico, China, Nicaragua (Alpha)	130,000	2,500
W17	2000	United States	FP	50% China, 50% US	120,000	1,200

(*continued*)

Table 11.3 Continued

Firm type[a]	Year est.[b]	Ownership[c]	Product type[d]	Fabric[e]	Product/ week[f]	Emp.[g]
W18	n.a.	United States	FP	15% US, 85% Nicaragua (Alpha)	65–70,000	1,600
W19	n.a.	Mexico, US	CMT/ FP	US, Mexico, Asia	100,000	1,600
W20	2008	United States	FP	United States, Nicaragua (Alpha)	100,000	2,000
W21	2007	United States	FP	50% China, 50% United States	105,600	800
W22	n.a.	United States	OBM	50% Pakistan and China, 50% US Mexico (<1%)	170,000	1,000–1,100
W23	2009	Mexico	Contract launderer	n.a.	200,000	1,100
W24	1999	Taiwan (China)	CMT	Depends on client, some Guatemalan	105,000	1,200
W25	2005	Taiwan (China)	FP	Asia	90,000	3,000
W26	2009	Trinidad	CMT/FP	United States, local (Alpha Textil), China	15,000	200
W27	2004	United States	CMT	Asia, US	50,000	1,000

Notes:
[a]K refers to knitwear manufacturers; W refers to wovens manufacturers.
[b]Year of establishment.
[c]Ownership of company.
[d]Production model: M denotes maquila; FP denotes full package; OBM denotes own-brand manufacturing.
[e]Origins of textiles.
[f]Units per week.
[g]Direct employment in owned and operated facilities.
Source: Firm interviews by authors.

in the United States or elsewhere in the hemisphere do not have manufacturing facilities outside of the Americas, several of the Asia-based companies boast production networks that are decidedly global in scope. This is particularly true for the Korean multinationals. For example, Sae-A Trading has 17 subsidiaries in six countries around the world, while Hansoll has 12 factories, principally in Asia (Viet Nam, Indonesia, Cambodia and the Philippines). Several of the US companies active in Nicaragua have owned and operated facilities elsewhere in the Americas (but not in Asia). One exception is the Nicaraguan subsidiary of VF Corporation, one of the world's largest blue jeans manufacturers, which is an outlier in the sample since VF owns many production facilities throughout Asia and Latin America.

A fourth and related finding pertains to the origins of the materials used in local production, and, therefore, the dependence of Asian versus American companies on the TPLs granted to Nicaragua. Although virtually every company interviewed used TPLs for at least some portion of their exports, the only companies that used TPLs for more than 50 per cent of their exports were those with Asian parent firms. In at least one case, all of the fabric used in Nicaragua was knitted in the company's owned-and-operated textile facility in China. The reliance of these firms on what may well prove to be short-term trade preferences, and the relative ease with which they can reorient orders across geographically extensive production networks, creates a serious challenge for economic and social upgrading in Nicaragua.

The challenges of trade-dependent development

Nicaragua confronts a number of difficulties as it attempts to parlay its recent export dynamism into a longer-term trajectory of economic and social upgrading. Its principal challenge lies in compensating for the loss of the trade preferences it has enjoyed under the CAFTA. Although Nicaragua continues to offer the lowest labour costs among the CAFTA countries, it cannot compete with countries such as Bangladesh and Viet Nam for the cheapest needle. Therefore, as both government officials and industry actors acknowledge, cheap labour will not sustain the country's position in the global apparel market.

At the same time, Asian exporters are able to exploit a regional textile base that is lacking in Central America. Although geographic proximity to the US market provides some advantage to manufacturers in the Western hemisphere, the global geography of textile production, centred in Asia, lessens this benefit. Minimizing the distance between the apparel and textile links in the chain (as opposed to the distance between the site of apparel assembly and the final market) implies a shorter supply chain for apparel companies, which in turn translates into lower transport costs, faster delivery times and potentially fewer bottlenecks and delays in the production process.

At the same time that Nicaragua is competing with Asia's exporters, it also faces a challenge from within its hemisphere. Haiti's fledgling export processing sector has received substantial investment in recent years from multiple sources, including the International Monetary Fund (IMF), the US government and the Inter-American Development Bank (Sontag 2012). Efforts to develop the country's export sector received increased attention after the 2010 earthquake, but initiatives to promote Haiti's apparel industry have been underway for years. Although not a signatory of CAFTA, Haiti received TPLs as part of a CAFTA side agreement in recognition of its status as the least developed country in the region. Up to 400 million square metre equivalents of non-originating apparel can enter the US market from Haiti each year (a TPL benefit four times greater than Nicaragua's). Haiti's TPLs are not only greater than Nicaragua's, they are also of longer duration, extending through 2018 instead of 2014 (Hornbeck 2010). The competition between Haiti and Nicaragua is heightened by the fact that, as mentioned already, Haiti is the only other country in the Americas to participate in the Better Work programme.

Knit apparel is the traditional mainstay of the Haitian apparel industry, accounting for more than three-fourths of the country's apparel exports to the United States. At the time of this fieldwork, one knitwear manufacturer, Sae-A, was in the process of inaugurating operations in Haiti. With 5,200 employees in five factories, Sae-A is one of the largest firms in Nicaragua. Yet plans call for the company's Haitian facilities to employ some 20,000 workers – more than three times the size of its Nicaraguan workforce (Sontag 2012). A few of the other companies interviewed were either actively pursuing investment opportunities in Haiti or considering doing so, largely motivated by a desire to retain access to TPLs once Nicaragua's TPL benefit expires. Among the companies actively considering such a move was a large manufacturer of jeans, whose presence in Haiti would further develop that country's modest but increasing supply of woven pants.

Virtually all firms acknowledged that the elimination of the TPLs would have a significant impact on their Nicaraguan operations. Some industry actors, presumably hoping to pressure policymakers into action, have made such statements publicly.[8] At the time of writing, officials in the Nicaraguan government were working with foreign buyers to lobby US policymakers for some extension of TPLs, possibly on a bilateral basis outside the auspices of CAFTA. However, it was unclear if these efforts will prove successful, and both public and private sector officials recognize that Nicaragua will eventually have to adjust to a post-TPL environment.

Thus, one of the major challenges confronting the industry – not just in Nicaragua, but in Latin America more generally – is developing a textile base that will enable exporters in the hemisphere to compete with Asia. To date, there is little evidence that the region's exporters are trying to coordinate their efforts in this regard. Although there is some regional supply of knit fabric in Central America, and Mexico manufactures a modest amount of denim, Nicaragua does not produce fabric. The country does have a large textile mill, however. US textile giant Cone Mills (part of the International Textile Group) built this facility, which opened in May 2008. It operated for less than two years and is currently closed, though the Nicaraguan government has been actively courting investors in the hopes that the plant may re-open. Aside from the Cone denim plant, the only other textile facility in Nicaragua is a converter that finishes imported greige goods (raw fabric that has not been dyed) as twill fabric.

But even if the textile base supporting Nicaragua's apparel manufacturers were strengthened, either domestically via investment in local fabric production or regionally via the purchase by Nicaraguan manufacturers of textiles being made elsewhere in the Americas, many apparel firms would continue to source some textiles from outside the region. Indeed, since in many cases it is the foreign buyer and not the local manufacturer that specifies the type and origin of fabric to be used in a particular order, this is not a decision over which apparel manufacturers necessarily have control.

Of course, if cost-competitive regionally produced textiles are available, manufacturers still have to be able to purchase them. Here one of the implications of the Americas-based versus Asia-based segments of the value chain is drawn out: most of the companies that were particularly concerned about financing were based in the Americas. These tended to be relatively small firms that had relocated production to Nicaragua, either from the United States or from elsewhere in Latin America (e.g. El Salvador or Mexico). Precisely because these companies lack an extensive network of owned-and-operated production facilities or global subcontractors in numerous countries, the Americas-based segment of the Nicaraguan apparel value chain may be less mobile than its Asia-based counterpart. In addition, while the Americas-based firms may be more closely tied to the future of Nicaragua's apparel industry than their Asian counterparts, their success may also be more contingent on broader factors such as access to cost-competitive textiles and the availability of credit, since they are less likely to be vertically integrated or have factories in Asia located close to that region's textile mills.

One notable development on the financing front is the extension to Nicaragua of the Global Trade Supplier Finance Programme (GTSF), which is a joint effort of Better Work and the IFC dating from 2010. This programme allows suppliers participating in Better Work to submit invoices for shipped goods to the Financing Programme, which then 'buys' the invoice in the form of an extremely low interest loan. This enables companies to manage the cash flow problems that result from the long delays between invoicing a buyer for delivered goods and receiving payment. Though currently it is unclear if a company's access to credit will be linked to its *performance* on the Better Work audit (as opposed to its *participation* in the programme), the Global Trade Supplier Financing Programme may offer the possibility for linking economic and social upgrading – a goal that has, until now, proven elusive.[9]

From factory compliance to social upgrading: the role of Better Work Nicaragua

This final section reviews the possibilities for Nicaragua to pursue upgrading via participation in the Better Work Nicaragua programme. The Nicaraguan government, particularly the CNZF, actively lobbied to secure inclusion in the programme, as did the US government. In October 2010, General Baltodano, the presidential delegate for investment and director of the CNZF, and US Secretary of Labor Hilda Solis announced a US$2 million grant to support Better Work Nicaragua.

During the authors' initial fieldwork in fall 2010, some scepticism was found among local manufacturers about the benefits of participating in Better Work. Some were worried that the programme's implementation in Nicaragua would send the wrong signal, causing the country to be perceived by foreign governments or global buyers as a 'problem case' in need of remedial attention. Because other Better Work countries lag in the area of protecting workers and enforcing labour laws, their concern was that Nicaragua's inclusion in the programme might indicate weakness rather than strength.

During the first year of the programme, Better Work Nicaragua staff worked hard to overcome these anxieties and enlist the participation of local firms, both through contacts with the local apparel industry association and by educating foreign buyers sourcing from Nicaragua about the project and encouraging them to secure the participation of their local suppliers. As of August 2012, Better Work Nicaragua had secured the participation of 11 factories, and was in the process of enrolling several more. Current plans call for the release of an initial Synthesis Report (an overall assessment of compliance at the country level) when at least 15 factories have been audited, since this would represent well over 25 per cent of the total number of garment factories (54, as of August 2012) currently operating in Nicaragua's free trade zones.

Almost half of the 11 companies that have agreed to participate in Better Work Nicaragua so far are Asian-based multinationals with plants in Nicaragua. This suggests that Asian-owned companies are overrepresented among Better Work participating factories, since Korean- or Taiwanese-owned companies account for a little more than one-third of the total number of apparel plants in the country (though they account for a greater proportion of apparel output and employment since they are among the largest firms in the industry). This may reflect their greater familiarity with the Better Work programme due to their participation in other countries where they have plants, such as Viet Nam and Indonesia, and it mirrors the uptake of the programme in Lesotho, as described by Pike and Godfrey in Chapter 8 of this volume. In contrast, American-owned companies have been relatively slow to enrol. However, Walmart recently announced that it will make participation in Better Work Nicaragua mandatory for its Nicaraguan suppliers.

The most distinctive feature of the Better Work programme vis-à-vis conventional compliance approaches is the technical and advisory services that staff members (known as enterprise advisers) provide participating factories. Following a Better Work audit, participating factories develop Performance Improvement Consultative Committees (PICCs). These bodies, comprised jointly of labour and management representatives, are charged with developing and implementing remediation measures to address whatever problems were identified in the audits. In this sense, Better Work has the potential for working with industry stakeholders, including foreign buyers, to develop an understanding of the factors creating downward pressure on wages and working conditions in global supply chains.

Better Work is premised on the hypothesis that there is, or can be, a positive feedback loop between economic and social upgrading. The advisory services that Better Work offers, and the support provided to PICCs at the enterprise level, are designed to identify causes of non-compliance and address them at the source. Improved performance on the audit is not only an indicator of social upgrading (e.g. insuring that wage payments and working conditions are consistent with the law, insuring a safe and discrimination-free work environment, etc.), but it can also be a means to economic upgrading as well. By identifying and addressing the root causes of violations, Better Work Nicaragua hopes to directly affect productivity, thus, facilitating economic upgrading as well. The factory-level data gathered in Nicaragua and in other Better Work countries provides a unique opportunity to test this hypothesis, and may yield evidence of a 'business case' for labour compliance that will incentivize companies to improve their performance. Thus, Better Work offers a unique opportunity to go beyond the standard approach to labour compliance represented by corporate code of conduct programmes, third party auditors or existing multi-stakeholder initiatives (see Anner 2012; Locke, Amengual and Mangla 2008; Locke and Romis 2010; Mayer and Gereffi 2010).

While the ultimate objective and design of the Better Work programme are consistent across all the participating countries, the criteria used to evaluate factories differ in each country. This is because the compliance assessment tool used by Better Work's auditors is based not only on the ILO's core labour standards but also on national labour laws. Consequently, Better Work's impact in a given country will depend on how much of an 'enforcement gap' exists between these standards and industry practices at the time of the programme's implementation. This presents a somewhat perverse challenge for Nicaragua, which is entering the programme at a relatively high baseline. As expected, the Better Work Nicaragua audits that had been carried out at the time of writing did not uncover any incidents of 'zero tolerance' violations (child labour, forced labour, etc.). Thus, demonstrating measurable improvement from this baseline will likely require progress on some of the issues that are more difficult to monitor and remediate.

One such example is freedom of association, which observers have noted is difficult to detect and remediate using conventional auditing methodologies (Anner 2012). In part, this is because the rights of workers to form unions and bargain collectively are likely to be more costly to employers and regarded as inimical to their interests, and the negative reputational consequences of non-compliance with workers' associational rights are perceived as lower than with other violations, such as child labour. However, freedom of association and collective bargaining are also more difficult to monitor and enforce because these are fundamentally dynamic *processes* rather than discrete *standards*, such as the number of fire extinguishers found on the factory floor or the hourly rate of overtime pay, which are, at least in theory, more easily verified during a particular audit (Rodríguez-Garavito 2005).

While evidence is still accumulating on Better Work Nicaragua, this project represents a challenge as well as an opportunity, not just in terms of achieving Better Work's aims of increasing labour compliance, but also in terms of Better Work's ability to advance the debate about labour outcomes in global industries beyond a narrow compliance perspective. The institutionalized social dialogue that exists in Nicaragua provides a unique infrastructure that Better Work Nicaragua can build on. For example, one of the first tasks necessary to get Better Work off the ground in any country is the formation of a Project Advisory Committee (PAC), which includes representatives of the three core constituencies of the ILO: government, workers and employers. Rather than beginning this process by identifying the appropriate stakeholders and convincing them to participate in such a structure, as Better Work staff in other countries have had to do, the Better Work Nicaragua staff was able to assemble a PAC by enlisting the participation of the groups that had negotiated the Tripartite Agreement. These PAC members played an important role in developing the Compliance Assessment Tool for Better Work Nicaragua's factory audits, thus securing a degree of stakeholder participation and input in the process that was unprecedented for the Better Work programme.

While initial indicators of Better Work Nicaragua's efforts are promising, we want to end on a note of caution. Better Work provides an opportunity to go beyond a cosmetic approach to labour compliance by trying to identify and address the root causes of violations. While some violations undoubtedly occur due to managerial incompetence or oversight, buyer practices also play a role. Therefore, Better Work should move beyond the tripartite structure of the ILO narrowly conceived and include the brands and retailers that are placing their orders with developing country exporters. These companies are not the direct employers of garment workers in Better Work countries, but their policies and practices – particularly with respect to lead times and pricing – critically shape the production process at the factory level and, therefore, directly affect workers (Anner, Bair and Blasi 2012). Thus, while it is necessary for local stakeholders, including the Nicaraguan government, trade unions and the employers' association, to pursue the opportunities for economic and social upgrading that Better Work provides, the ultimate success of the project may be contingent on the meaningful participation of the brands that are the clients of Nicaragua's manufacturers.

The importance of global buyer participation was repeatedly underscored in the authors' interviews. Specifically, company representatives and industry officials expressed a strong desire to see a more active commitment to Better Work on the part of foreign apparel buyers. While local manufacturers were encouraged that some buyers promised to reduce or eliminate their audits of local firms participating in Better Work, they are hopeful that buyers might provide more incentives for local factories to participate (e.g. by pledging to source only from Better Work factories). One informant expressed his view this way:

> The brands have to understand that they need to be involved in this process. If it [Better Work] doesn't generate contracts for companies, no one is going to participate because it won't have any value added for them.

To date, the only concrete commitment made by brands is that they will reduce or eliminate the audits they would otherwise carry out in participating factories, meaning they will accept the results of the Better Work assessments in place of their own. This is far short of what the factories would like to see. While purchasing guarantees or other kinds of concrete commitments on the part of the brands will not be easy to achieve, it is critical that buyers do more than simply encourage their suppliers to participate. They need to be full and willing partners in a dialogue with suppliers in order to identify, and ultimately alter, buyer practices that contribute to non-compliance and inhibit more meaningful forms of social upgrading. Combined economic and social upgrading has the best chance of benefitting local workers and their families, not just in Nicaragua but all along the apparel value chain.

Notes

1. The signatories of CAFTA – the United States, Costa Rica, Dominican Republic, Honduras, Guatemala, El Salvador and Nicaragua – ratified and implemented the treaty individually, which meant that it became operative in different countries at different times. In Nicaragua, CAFTA entered into force in April 2006.
2. The authors' research was part of a study commissioned by the Nicaraguan government, and specifically by the Secretariat of the National Free Zones Commission (CNZF). Officials at CNZF wanted a diagnostic study of the strengths and weaknesses characterizing the Nicaraguan apparel industry, and the prospects for improving its competitiveness, particularly in the context of the CAFTA-DR trade agreement with the United States. Through resources provided by the US Agency for International Development (USAID) and its local programme 'Nicaragua Empresas y Empleo' (carried out by CARANA), the Center on Globalization, Governance & Competitiveness (CGGC) at Duke University in Durham, North Carolina was commissioned to carry out this study in fall 2010 (Gereffi and Bair 2010). Ingrid Veronica Mujica and Stacey Frederick of CGGC contributed to the initial research and report. This chapter also draws from follow-up fieldwork in both Nicaragua and Honduras that was conducted by one of the authors in summer 2011, as well as subsequent updates from informants in summer 2012.

3. In addition, three other non-apparel companies within the free trade zone sector were interviewed: a call centre, a furniture manufacturer and a company producing disposable medical equipment.
4. Knit apparel accounts for a greater percentage of Nicaragua's total garment exports. In 2009, the value of knit apparel exports to the US market (US$4.9 billion) was more than three times the value of woven apparel exports (US$1.4 billion).
5. Although the fabric these companies use is produced in another CAFTA country (Honduras), the shirts sewn in Nicaragua do not necessarily qualify as CAFTA-originating because some of the yarn used in the Honduran mills comes from outside the region. For this reason, companies using this regional production model may still require TPLs for some portion of their production.
6. The Tripartite Agreement model dates from 2009, when dislocations in the export sector caused by the US recession and the consequent decline in apparel orders from foreign buyers led government officials to propose and negotiate with private sector and labour leaders an Emergency Economic and Labor Agreement in March 2009. This Agreement created the Free Zone Tripartite Labor Commission as a forum for dialogue and cooperation between the parties, with the goal of strengthening the industry and preserving jobs in the textile and apparel sector. It also established specific minimum wage increases for 2009 and 2010 (8 and 12 per cent increases, respectively). In January 2010, the same parties signed a new Agreement that outlined a schedule of minimum wage increases from 2011 through 2013 (8 per cent in 2011, 9 per cent in 2012 and 10 per cent in 2013).
7. The widespread perception is that the government was motivated to pursue the Agreement by a wave of job losses in the industry, most notably those precipitated by the decision of the large Taiwan-based multinational apparel manufacturer Nien Hsing to abandon its sewing operations in Nicaragua in 2008, leading to a loss of some 14,800 jobs.
8. For example, Randy Price, vice president of manufacturing for VF Corporation (a company which produces jeans and khaki pants in an owned and operated facility in Nicaragua), has stated that the TPL issue is 'imperative to VF's Western Hemisphere strategy'. Furthermore, because 'a company as large as VF needs 12 to 18 months to plan production and raw material input strategies', the parent corporation will make a decision about whether to maintain production volumes in Nicaragua well in advance of the 2014 expiration date, adding significant pressure and making the timely resolution of this issue critical (Nichols 2011).
9. The IFC pioneered a programme that links loan conditions to investments that enable environmental upgrading (the Climate Smart Trade initiative); this provides a precedent for incentivizing social upgrading among suppliers participating in Better Work.

Conclusions

Arianna Rossi, Amy Luinstra and John Pickles

The chapters in this volume have demonstrated in various ways that the globalization of the apparel industry, and its increasing organization through global value chains, has enormous consequences for the kind of jobs and conditions of work being produced in different countries. This concluding chapter summarizes some key findings about the quality of jobs and on-going challenges facing decent work, the changing roles of private and public governance, and the specific effects of the Better Work programme. Policy recommendations emerging from the chapters are outlined and some directions for future research in this field are identified.

Global crisis and opportunity

The fragmentation of apparel value chains has been fostered by *defensive* trade policies in major markets: textile and apparel manufacturers in industrialized countries have fought hard to retain their primary markets and, as the quota system and the subsequent further liberalization of trade encouraged the outsourcing and off-shoring of apparel assembly, they mobilized politically to ensure that US and EU textile manufacturers retained their initial advantages and remained the primary input suppliers for the sewing operations that were being relocated off-shore.

Whether coordinated by brands, retailers or trading intermediaries, global value chains thus both degrade work and improve it. They exacerbate the challenge of regulating for decent work and provide new opportunities for different kinds of work and different forms of regulation. They are also largely responsible for driving the 'race to the bottom' and are constantly innovating new terms of engagement with workers around the world. In some cases, workers themselves have seen their positional power in global value chains shift from one of low-cost interchangeable hands for hire to variously valued partners in managing broader competitive and efficiency pressures. While suppliers in Africa and Latin America struggle to gain orders and suppliers in Bangladesh struggle to expand factories to meet orders, lead firms in value chains recognize that at this time, at least, improving work and working conditions in their supply chains are essential to their own economic security.

If the hyper-exploitation model of deregulated hands-off global contracting since the 1970s has degraded working conditions and wages, it now generates reputational risk and economic costs for lead-firms. Costly production delays, higher return rates and negative shareholder value are all at risk in poorly regulated factory work. As wage and logistics costs increase, such inefficiencies in the labour process compound the risks faced. As the writing of this volume comes to completion (April 2013), the world's media is focused on the Rana Plaza factory collapse in Dhaka where 1,129 people were crushed to death after having been forced back to work despite a call to evacuate the previous day as cracks in the building were detected. Like many Bangladeshi factory owners rushing to capture the contracts flooding into the country, the owner of this establishment has added two extra floors beyond those for which he had sought and been given permission. In the immediate aftermath, major international buyers were rushing to check their position in the factory. Were they sourcing in the factory? Had they signalled any initial plans to source? What was their liability? These same lead firms that had rushed to new low-cost sourcing frontier after frontier over the past two decades are now forced to take much more seriously their suppliers' compliance with local construction laws and in many instances to push for better practices than those laws allow for, even if they were enforced.

The result has been a complex series of quantitative limits, safeguard actions, rules of origin, duties and tariffs on imports exercised often differentially by the importing countries at the product and exporting country level. One consequence has been the distribution of low-value assembly work across an increasingly wide range of countries. With the end of quotas, apparel sourcing shifted rapidly with some countries gaining orders and expanding employment, while other countries saw their quota-based industry decline and jobs were lost. China, in particular, emerged as the largest exporter of clothing, with 37 per cent of the global total in 2011, but export production and jobs also increased rapidly in Bangladesh, Indonesia and Viet Nam. In this process, prices and wages were driven down across the industry, setting a new norm that for many was below living wage levels.

In turn, the resulting *competitive* pressures and race to the bottom generated their own limits. Pressure from workers, consumers and buyers for economic and social upgrading increased across the chain, and – whether to manage business risk in high-value consumer markets or to protect worker safety in second and third tier subcontracting factories – stakeholders in the industry have struggled to create an institutional environment, including for example, conditionalities in trade agreements, to ensure that claims by workers for higher returns to wages, larger investments in social wages and investments in workplace safety (social upgrading) can be achieved in sustainable fashion provided they are linked to parallel increases in productivity (economic upgrading).

The result is a complex landscape of economic and social upgrading and downgrading, in which the fortunes of manufacturers and workers in low-income countries have varied greatly. Producers in countries such as China, India, Sri Lanka and Turkey are upgrading into higher-value segments of the value chain by adding new functions (such as branding and design), diversifying their markets into regional and domestic markets, and creating either backward or horizontal linkages in the production process. Such suppliers rely increasingly on skills upgrading and training, but these are achievable only by increases in direct wages, social wage payments and/or increased rights for workers. Comparable developments in other manufacturing centres and competition from the service sector are also driving up the demand for up-skilling and higher wages as regional labour markets tighten.

Producers in countries that have not been able to upgrade production systems or capture higher value segments of the market, such as Morocco, are adapting their production to buyer demands for fast turn and fast fashion. The results for these countries are mixed. Contracts have expanded and employment has increased (at least until the effects of the 2008 recession were felt). But workers were exposed to production line reorganization aimed at speed and flexibility. Their benefits were restricted to more stable jobs and higher workplace and product standards, but at low levels of pay and under increasingly tight production deadlines. In other countries, such as Jordan, preferential market access agreements such as qualified industrial zones (QIZs) lock suppliers into low-wage assembly production using imported fabric and migrant workers, generating few if any backwards linkages, disappointing employment effects and little spill-over effects into the local economy.

Producers in former quota-driven export platforms, such as Kenya, Lesotho and Madagascar, continue to struggle to sustain markets and have become increasingly dependent on preferential access to major markets (such as AGOA). The uncertain status associated with the renewal of these programmes has had negative effects. For the most part, workers in these factories face uncertain employment, low wages and often despotic working conditions where non-payment of wages, forced overtime and hazardous, and at times harassing, work environments prevail.

The prominence of China as product production platform for many lead firms has also emerged as a weakness. Many lead firms continue to source the majority of products from China, but they also seek to diversify into other countries. The rise of strategic partnerships between buyers and suppliers is also part of a broader recalibration of sourcing logics as production costs and future uncertainties increase in China. Making use of multiple supply chains, balancing global sourcing from Asia with quick turn local production with regionally proximate production has now become the norm for major brands and retailers, offering them both a greater level of supply stability along with reserve capacity to meet unexpected demand.

At the same time, the needs of retailers for volume, quality, flexibility and timing have become more important and have changed the power dynamics in global value chains, allowing suppliers to take on many more functions and operate with more freedom than was the case even a few years ago. Thus, as supply chains have become more footloose and flexible for some companies, for other lead firms they have become more concentrated and focused on strategic alliances and partnerships.

Job quality

If jobs in apparel global value chains are to actually contribute to worker and community development and empowerment, it is critical to look at both quality and scale of employment. While Bernhardt shows that there have been almost no cases in which social upgrading has occurred without economic upgrading (Chapter 2), he also confirms that there are many instances where economic upgrading does not translate into social upgrading.

Going further in the analysis of quality of jobs, Vaughan-Whitehead concludes, in Chapter 3, that wages in the global apparel value chains remain low and often below legally mandated minima, as also illustrated by Miller in Chapter 4. Both of these authors call for effective worker participation in any effort to negotiate improvement in wages or working conditions. Where trade unions do not exist, there remains a need for workers to be allowed to organize freely to discuss the allocation of such wages and supplements. Miller, in particular, points to the crucial role international framework agreements, multi-stakeholder initiatives and individual firm decisions (such as Knights Apparel's living wage commitment) can play in establishing negotiated improvements in working conditions and economic performance. In the absence of such kinds of commitments and agreements, Vaughan-Whitehead stresses the effects of high turnover in the industry, a finding observed also by Record, Kuttner and Phouxay in Chapter 6 and by Bair and Gereffi in Chapter 11, which can be interpreted as a symptom of workers' dissatisfactions with working conditions.

The authors in this volume have called for a large variety of policy interventions in their contributions, based on their analytical and empirical findings. Emphasis has been given to the role of public governance actors, and in particular to the role of the state and its apparatus, including labour administration and inspection. Vaughan-Whitehead (Chapter 3) argues that it is paramount to strengthen compliance with wage regulations through effective enforcement in order to be able to push the fair wage agenda. The lack of enforcement capacity is also highlighted in Pike and Godfrey's chapter on the Lesotho case (Chapter 8), arguing that compliance with low standards is not sufficient to ensure workers' wellbeing. From a different viewpoint, state institutions are shown to be playing a key role in mediating commercial pressures in global value chains and in providing opportunities for social upgrading in the examples of Eastern European countries discussed by Mayer and Pickles in Chapter 1 and Plank, Rossi and Staritz in Chapter 5. It is also suggested that the government has a role to play in stimulating systemic competitiveness based on skill development, economic and social upgrading strategies, as noted by Plank, Rossi and Staritz (Chapter 5), Pike and Godfrey (Chapter 8) and Bair and Gereffi (Chapter 11).

The role of private governance

In terms of private governance, the role of global buyers as catalysts for sustainable governance of global value chains has been highlighted in various occasions throughout this collection, but with a significant caveat: buyers must align their purchasing practices with labour standards requirements. Miller (Chapter 4), Plank, Rossi and Staritz (Chapter 5) and Pike and Godfrey (Chapter 8) show that a change in buyers' decision making, reconciling the tension between their sourcing practices and their corporate social responsibility strategies, would have a direct and positive impact on social upgrading and workers' wellbeing. Bair and Gereffi, in Chapter 11, also call for a root-cause analysis of non-compliance in the Nicaraguan apparel industry, especially identifying those issues that may be caused by buyers' purchasing practices. Related to this, Pike and Godfrey recommend buyers to promote growth in their supply base by maintaining longer term relationships with their suppliers (Chapter 8). In Chapter 9, Arnold urges buyers to eliminate the misuse of fixed duration contracts.

Highlighting another important aspect of the private sector's role in improving working conditions in supplier factories, Brown, Dehejia and Robertson recommend, in Chapter 10, thinking about human resource innovations, and specifically a configuration that is compliant with labour standards, as another form of technology that might improve overall factory performance. Their analysis shows that factories very rarely decide to revert into a non-compliant behaviour once they have made the initial decision to be compliant, suggesting that they are not observing an increase in cost, but rather mostly an increase in productivity as a result of improved working conditions.

In their work on fast fashion, Plank, Rossi and Staritz also show how improvements in working conditions and productivity have emerged as a result of the growth of fast fashion, quick-turn production systems and competition for declining markets (Chapter 5). But they have also shown how pressure on cost and time results in the need to create different kinds of workers. Factory owners in Morocco have managed this tension by stratifying permanent and temporary or precarious workers, with differential rewards and working conditions for each. Elsewhere, these pressures have created very different outcomes. In Romania, retaining a core workforce has resulted in the intensification of outsourcing to less compliant irregular factories. In Jordan, sourcing expansion under the US QIZ trade agreement has led to similar pressures on cost and time, but in this case, instead of segmentation of permanent and temporary contract workers or increased secondary sub-contracting the industry has responded with the wholesale importation of South Asian migrant temporary contract workers to the export processing zones.

The work of Better Work

In Part II, several chapters analysed from different perspectives the Better Work programme – Pike and Godfrey in Lesotho (Chapter 8), Arnold (Chapter 9) and Brown, Dehejia and Robertson (Chapter 10) in Cambodia and Bair and Gereffi (Chapter 11) in Nicaragua. These contributions did not directly aim at assessing the impact of the programme on workers' lives and on competitiveness (partly due to the recent establishment of the programmes and the limited data availability), but they do indicate that there is an emerging consensus and political dynamic to mobilize the potential of whatever more synergistic forms of governance might be possible. Whether at the global level (in the form of buyers), at the national and sectoral level (involving national governments, employers' organizations and trade unions) down to the factory level (directly engaging with factory managers and workers), the marginal distribution of surplus to wages and the social, political and economic benefits of Better Work are now widely recognized.

It may well be the case that these opportunities have arisen because of the crises and costs generated by predatory business practices, low-road development strategies and the squeezing of the conditions and price of labour. It is certainly clear to many of our authors that best practice models of workplace reorganization also function as part of lead-firm strategies of market positioning and competitive advantage. Nonetheless, interest in better working conditions and employment practices has been widely acknowledged and coordination efforts from different actors have expanded. Their ability to convene stakeholders with often conflicting interests has emerged as key characteristics in efforts to design more sustainable governance scenarios in the global apparel value chain.

In Chapter 9, Arnold points to shortcomings in Better Factories Cambodia as an instrument for improving industrial relations, based on perspectives from the fractured trade union movement in Cambodia and highlighting the constraints for the full expression of freedom of association. He argues that the results from Cambodia have generated uncertain outcomes for independent trade unions. The limitations that he points to, such as the lack of monitoring of working conditions in subcontracting factories, concern the complexity of global value chains themselves and the challenge of reaching the most vulnerable actors. While the industrial relations environment in Cambodia is challenging and ever-changing, and national policies and Better Factories Cambodia may have contributed to the proliferation of the number of unions in individual factories, compliance with formal indicators of freedom of association and industrial relations, such as the presence of shop stewards and union liaison officers in factories, does seem to have improved since the establishment of Better Factories Cambodia (Rossi and Robertson 2011).

At the factory level, Miller recommends, in Chapter 4, that Better Work's training for workers include how to negotiate effectively with management in production quota setting, drawing on existing trade union expertise in this area. In Chapter 8, Pike and Godfrey also suggest in the Lesotho case that Better Work develop strategies for ongoing worker involvement, sustainable training programmes and building union capacity. At the industry and value chain level, Bair and Gereffi argue in Chapter 11 that Nicaragua could be an opportunity for Better Work to pursue 'second generation' compliance issues, including those more systemic to the functioning of the global apparel value chain, given the comparatively bright prospects for social dialogue in the industry.

Directions for future research

The volume's contributions call for further research on a number of aspects. A key dimension that needs further investigation is the direction of causality between economic and social upgrading trends, as discussed by Bernhardt in Chapter 2 and Brown, Dehejia and Robertson in Chapter 10. Evidence has been presented supporting both the hypotheses that economic upgrading is an indispensable condition for social upgrading and that improved productivity and business performance result from improved working conditions.

Improvement in both labour standards and productivity are objectives that, at least to some extent and under some conditions, are compatible. Efforts towards social upgrading have focused on improvements in working and living conditions, strengthening worker representation and dispute resolution mechanisms within factories and across the industry, as well as on modernization of human resource and production management systems. These efforts could be significantly enhanced by longer term investment in building a skilled and healthy workforce through enhanced public–private partnerships and programmes. There is also evidence, shown by Brown, Dehejia and Robertson in Chapter 10, that suppliers' decision to become compliant largely depends on the adjustment costs associated with it. The provision of incentives to make these investments, such as through the Better Work programme, has allowed some factories to 'take the leap' and discover later that the changes were indeed beneficial. As there is growing evidence of irreversibility effects in business practices, initial investment incentives to encourage compliance would be a good strategy to follow; subsidizing one-time investments and sharing information about the specific reforms that factories have tried and have proven to be beneficial would help other factories overcome obstacles and lead to more beneficial improvements in working conditions.

Fully understanding the causal links between economic and social upgrading, including the role of global buyers in fostering or hampering these two dimensions, is crucial to the design of effective policies. Subsequently, a deeper understanding of the role of different national policies (industrial policies, trade policies, labour market policies) is needed to appreciate what coherent configuration can advance workers' livelihoods and wellbeing while at the same time ensuring the promotion of competitiveness in global apparel value chains – in other words, to ensure that economic and social upgrading can occur simultaneously and reinforce each other.

Strengthening the findings and the recommendations presented in this volume with regard to private governance, further research is needed to explore how global buyers can govern their supply chains in ways that ensure fair wages, safe working conditions and respect for workers' rights while staying competitive and innovative. A deeper look at purchasing practices will also enable targeted policy recommendations. In particular, it is clear that lead firms in apparel global value chains are currently experiencing a major change in outlook in regard to who and what ought to be regulated. Private governance models are perhaps more well-developed for apparel than for almost any other sector; but significant limits to these models are now acknowledged even by lead-firms themselves, who are now planning for post-monitoring forms of regulation. While state regulation was greatly undermined by the expansion of global value chains, lead firms struggling with 'code overload' and 'monitoring fatigue' are now asking national and local administrations to be more effective partners in maintaining standards. With the decline of neoliberal orthodoxy, governments are increasingly concluding that their desire for economic upgrading need not come at the expense of abandoning their regulatory and social protection functions. Indeed, some states are increasingly recognizing that workplace standards and enforcement mechanisms can be essential elements of value chain upgrading. The strengthening of state governance capacities, especially in emerging economies like China, India and Brazil, are potential game-changers.

The evidence and analysis presented in this volume show that global value chains may be the driver of the race to the bottom in terms of labour standards, but may also be emerging as a driver of social upgrading, particularly in the apparel industry. The outstanding question is, therefore, not how to escape the pressures generated by global outsourced production, but rather how to harness the potential of global value chains for change that is sustainable both from the perspective of firms and workers.

Bibliography

Abernathy, F. H., Dunlop, J. T., Hammond, J. H. and Weil, D. (1999). *A Stitch in Time: Lean Retailing and the Transformation of Manufacturing – Lessons from the Apparel and Textile Industries* (Oxford: Oxford University Press).

Abernathy, F. H., Volpe, A. and Weil, D. (2006). 'The Future of the Apparel and Textile Industries: Prospects and Choices for Public and Private Actors'. *Environment and Planning A* 38(12), 2207–32.

Acona (2004). *Buying Your Way into Trouble? The Challenge of Responsible Supply Chain Management* (London: Insight Investment Management Limited).

Adams, B. (2012). '10,000 Days of Hun Sen'. *The New York Times*, 31 May. Available at: http://www.nytimes.com/2012/06/01/opinion/10000-days-of-hun-sen.html (accessed 13 August 2013).

Adler, D. and Woolcock, M. (2010). 'Justice without the Rule of Law? The Challenge of Rights-Based Industrial Relations in Contemporary Cambodia'. In C. Fenwick and T. Novitz (eds) *Human Rights at Work: Perspectives on Law and Regulation* (Oxford: Hart Publishing).

Aiginger, K. (1997). 'The Use of Unit Values to Discriminate between Price and Quality Competition'. *Cambridge Journal of Economics*, 21(5), 571–92.

Aiginger, K. (1998). 'A framework for evaluating the dynamic competitiveness of countries'. *Structural Change and Economic Dynamics*, 9(2), 159–88.

American Center for International Labor Solidarity (ACILS); Cambodian Center for Human Rights; Cambodian Center for Independent Media; Cambodian Human Rights Action Committee; Community Legal Education Center; Housing Rights Task Force; Cambodian League for the Promotion and Defense of Human Rights. (2010). 'Joint media statement: Intimidation and legal threats against union workers and leaders must cease', September, unpublished.

Amighini, A. (2006). 'Upgrading in International Trade: Methods and Evidence from Selected Sectors'. In C. Pietrobelli and R. Rabellotti (eds), *Upgrading to Compete: Global Value Chains, Clusters, and SMEs in Latin America* (Washington, DC: Inter-American Development Bank) pp. 221–50.

Anas, A. Z. M. (2011). 'Garment Wage Hike Deprives Sr Workers: ILO Deputy Chief'. *The Financial Express*, 17 January. Available at: http://www.thefinancialexpress-bd.com/more.php?news_id=123162&date=2011-01-17 (accessed 12 July 2012).

Anderson, S. T., Friedman, D. and Oprea, R. (2010). 'Preemption Games: Theory and Experiment'. *American Economic Review* 100(4), 1778–803.

Ang, D., Brown, D., Dehejia, R. and Robertson, R (forthcoming). 'Public Disclosure, Reputation Sensitivity, and Labor Law Compliance: Evidence from Better Factories Cambodia'. *Review of Development Economics.*

Anner, M. (2012). 'Corporate Social Responsibility and Freedom of Association Rights: The Precarious Quest for Legitimacy and Control in Global Supply Chains'. *Politics & Society* 40, 609–44.

Anner, M., Bair, J. and Blasi, J. (2012). *Buyer Power, Pricing Practices, and Labor Outcomes in Global Supply Chains*. Institute of Behavioral Science Institutions Working Paper INST2012-11.

Antony, P. and Gayathri, V. (2002). 'Child Labour: A Perspective of Locale and Context'. *Economic and Political Weekly*, 28 December.

Appelbaum, R. P. (2004). 'Fighting Sweatshops: Problems of Enforcing Global Labour Standards'. In R. P. Appelbaum and W. I. Robinson (eds), *Critical Globalization Studies* (New York: Routledge) pp. 369–78.

Appelbaum, R. P., Bonacich, E. and Quan, K. (2005). *The End of Apparel Quotas: A Faster Race to the Bottom?* University of California, Santa Barbara: Global & International Studies Program Working Paper 35. Available at: http://repositories.cdlib.org/gis/35 (accessed 20 September 2011).

Arnold, D. and Pickles, J. (2011). 'Global Work, Surplus Labor, and the Precarious Economies of the Border'. *Antipode*, 43(5), 1598–624.

Arnold, D. and Toh, H. S. (2010). 'A Fair Model of Globalisation? Labour and Global Production in Cambodia'. *Journal of Contemporary Asia* 40(3), 401–24.

Asda George (2012). *Doing the Right Thing Update 2012–2013*. Available at: http://your.asda.com/system/dragonfly/production/2013/03/08/10_24_03_786_George_DTRT_Report_2012_2013_SPREAD_.pdf (accessed 16 September 2013).

Asociación de Investigación y Estudios Sociales (ASIES) (2007). VI Encuesta a la industria de vestuarios y textiles, July (Guatemala: Konrad Adenauer Stiftung).

Asociación de Investigación y Estudios Sociales (ASIES) (2010). Estudio de opinión de los empresarios de vestuario y textiles sobre el comportamiento de variables económicas y financieras de su actividad productive, August (Guatemala: Konrad Adenauer Stiftung).

Asuyama, Y., Chhun, D., Fukunishi, T., Neou, S. and Yamagata, T. (2010). *Firm Dynamics in the Cambodian Garment Industry: Firm Turnover, Productivity Growth, and Wage Profile under Trade Liberalization*. Institute for Developing Economies Discussion Paper No. 268, December. Available at: https://ir.ide.go.jp/dspace/bitstream/2344/930/1/ARRIDE_Discussion_No.268_asuyama.pdf (accessed 13 August 2013).

Baccaro, L. (2001). *Civil Society, NGOs, and Decent Work Policies: Sorting Out the Issues*. International Institute of Labour Studies Discussion Paper No. 127 (Geneva: ILO).

Bachman, S. (2000). 'A New Economics of Child Labor: Searching for Answers Behind the Headlines'. *Journal of International Affairs* 53(2), 545–72.

Baden, S. (2002). *Trade Policy, Retail Markets and Value Chain Restructuring in the EU Clothing Sector*. PRUS Working Paper No. 9 (Poverty Research Unit at Sussex. Brighton: University of Sussex).

Bair, J. (2008). 'Analyzing Global Economic Organization: Embedded Networks and Global Chains Compared'. *Economy and Society* 37(3), 339–64.

Bair, J. (2009). Global Commodity Chains—Genealogy and Review. In J. Bair (ed.), *Frontiers of commodity chain research* (Stanford: Stanford University Press).

Bair, J., Dickson, M. and Miller, D. (eds) (2013). *Workers' Rights and Labor Compliance in Global Supply Chains: Is a Social Label the Answer?* (London: Routledge).

Bair, J. and Werner, M. (2011). 'The Place of Disarticulations: Global Commodity Production in La Laguna, Mexico'. *Environment and Planning A* 43(5), 998–1015.

Baldwin, R. (ed.) (2009). *The Great Trade Collapse: Causes, Consequences and Prospects*. A VoxEU.org Ebook. Available at: http://voxeu.org/index.php?q=node/4297 (accessed 13 August 2013).

Barrientos, S. (2013). '"Labour Chains": Analysing the Role of Contract Workers in Global Production Networks'. *Journal of Development Studies* 49(8), 1058–71.

Barrientos, S., Gereffi, G. and Rossi, A. (2011). 'Economic and Social Upgrading in Global Production Networks: A New Paradigm for a Changing World'. *International Labour Review* 150(3–4), 319–40.

Barrientos, S., Mathur, M. and Sood, A. (2010). *Decent Work in Global Production Networks: Challenges for Vulnerable Workers in the Indian Garments Sector*. In A. Posthuma and D. Nathan (eds), *Labour in Global Production Networks in India*. (Oxford: Oxford University Press), pp. 126–43.

Barrientos, S. and Smith, S. (2007). 'Do Workers Benefit from Ethical Trade? Assessing Codes of Labour Practice in Global Production Systems'. *Third World Quarterly* 28(4), 713–29.

Bazeley, P. (2002). 'The Evolution of a Project Involving an Integrated Analysis of Structured Qualitative and Quantitative Data: From N3 to NVivo'. *International Journal of Social Research Methodology* 5(3), 229–43.

Begg, R., Pickles J. and Smith, A. (2003). 'Cutting It: European Integration, Trade Regimes, and the Reconfiguration of Eastern and Central European Apparel Production'. *Environment and Planning A* 35(12), 2191–207.

Behrend, H. (1957). 'The Effort Bargain'. *Industrial and Labor Relations Review* 10(4), 503–15

Beresford, M. (2009). 'The Cambodian Clothing Industry in the post-MFA Environment: A Review of Developments'. *Journal of the Asia Pacific Economy* 14(4), 366–88.

Berik, G., and van der Meulen Rodgers, Y. (2010). 'Options for Enforcing Labour Standards: Lessons from Bangladesh and Cambodia'. *Journal of International Development* 22, 56–85.

Bernhardt, T. (2013). *Developing Countries in the Global Apparel Value Chain: A Tale of Upgrading and Downgrading Experiences*. Capturing the Gains Working Paper 2013/22 (Manchester: Brooks World Poverty Institute, University of Manchester).

Bernhardt, A., Milkman, R., Theodore, N., Heckathorn, D., Auer, M., DeFilippis, J., Luz González, A., Narro, V., Perelshteyn, J., Polson, D. and Spiller, M. (2008). *Violations of Employment and Labor Laws in America's Cities, Broken Laws, Unprotected Workers* (Chicago: Center for Urban Economic Development; New York: National Employment Project; Los Angeles: UCLA Institute for Research on Labour and Employment).

Bernhardt, T. and Milberg, W. (2013). 'Does Industrial Upgrading Generate Employment and Wage Gains?' In A. Bardhan, D. Jaffee, and C. Kroll (eds), *The Oxford Handbook of Offshoring and Global Employment* (New York: Oxford University Press) pp. 490–533.

Better Factories Cambodia (2006a). *Cambodian Garment Industry: One year later* (Geneva: ILO). Available at: http://www.ilo.org/asia/info/public/pr/WCMS_BK_PR_162_EN/lang--en/index.htm (accessed September 2013).

Better Factories Cambodia (2006b). *Better Factories Concerned about Labour Contracts*, Quarterly Newsletter No. 6, October. Available at: http://betterfactories.org/?p=3272 (accessed September 2013).

Better Factories Cambodia (2009). Newsletter No. 13, May. Available at: http://betterfactories.org/?p=3295 (accessed 16 September 2013).

Better Factories Cambodia (2010). *26th Synthesis Report on Working Conditions in Cambodia's Garment Sector* (Phnom Penh: ILO).

Better Factories Cambodia (2012). *28th Synthesis Report on Working Conditions in Cambodia's Garment Sector* (Phnom Penh: ILO).

Better Work (2011). Programme Update March April 2011. Available at: http://betterwork.org/global/?p=1004 (accessed September 2013).

Better Work Haiti (2010) *Garment Industry: 1st Biannual Report under the HOPE II Legislation* (Geneva and Washington, DC: ILO and IFC).

Better Work Jordan (2010). *Garment Industry 1st Compliance Synthesis Report* (Geneva and Washington: ILO and IFC).

Better Work Jordan (2011). *Garment Industry 2nd Compliance Synthesis Report* (Geneva and Washington, DC: ILO and IFC).

Bhaskaran, R., Nathan, D., Phillips, N. and Upendranadh, C. (2010). 'Home-based Child Labour in Delhi's Garments Sector: Contemporary Forms of Unfree Labour in Global Production'. *Indian Journal of Labour Economics* 53(4), 607–24.

Bheda, R. (2002). 'Productivity in the Indian Apparel Industry: Paradigms and Paragons'. *Journal of Textile and Apparel, Technology and Management* 2(3), 1–9.

Bickham Mendez, J. (2005). *From the Revolution to the Maquiladoras: Gender, Labor and Globalization in Nicaragua* (Durham, NC: Duke University Press).

Biørn, E., Golombek, R. and Raknerud, A. (1998). 'Environmental Regulations and Plant Exit'. *Environmental and Resource Economics* 11, 35–59.

Birnbaum, D. (2008). *Crisis in the 21st Century Garment Industry and Breakthrough Unified Strategies* (New York: FashionIndex).

Bloom, N., Bond, S. and Van Reenen, J. (2007). 'Uncertainty and Investment Dynamics'. *Review of Economic Studies* 74, 391–415.

Bobo, K. (2008). *Wage Theft in America: Why Millions of Working Americans Are Not Getting Paid – And What We Can Do About It* (New York: New Press).

Boggis, J. J. (2001). 'The Eradication of Leisure'. *New Technology, Work and Employment* 16(2), 118–29.

Breman, J. (2010). *Outcast Labour in Asia: Circulation and Informalization of the Workforce at the Bottom of the Economy* (New Delhi: Oxford University Press).

Brooks, E. C. B. (2007). *Unravelling the Garment Industry: Transnational Organizing and Women's Work* (Minneapolis, MN: University of Minnesota Press).

Brown, D. and Deardoff, A. (2011). 'The Apparel Industry and the Jordanian Economy. Calculating the Domestic Share of Sector Value Added'. Report prepared for Better Work Jordan, November 2011.

Brown, D., Dehejia, R. and Robertson, R. (2011a). *Is There a Business Case for Improving Labor Standards? Some Evidence from Better Factories Cambodia.* Working paper, mimeo, Macalester College.

Brown, D., Dehejia, R. and Robertson, R. (2011b). *Working Conditions and Factory Survival: Evidence from Better Factories Cambodia.* Better Work Discussion Paper Series No. 4 (Geneva and Washington, DC: ILO and IFC).

Bulgun, E. and Vuruskan, A. (2005). 'Development of a Software about Calculating the Production Parameters in Knitted Garment Plants'. *World Academy of Science, Engineering and Technology* 9, 6–13.

Burawoy, M. (2010a). 'From Polanyi to Pollyanna: The False Optimism of Global Labor Studies'. *Global Labour Journal* 1(2), 301–313.

Burawoy, M. (2010b). 'On Uncompromising Pessimism: Response to my Critics'. *Global Labour Journal* 2(1), 73–77.

Burra, N. (2005). 'Crusading for Children in India's Informal Economy'. *Economic and Political Weekly*, 3 December.

Buttle, M. (2007). Inditex and the ITGLWF Sign Framework Agreement. Impactt.

Cambodia Institute of Development Study (CIDS) (2009a). *Living Wage Survey for Cambodia's Garment Industry* (Cambodia Institute of Development Study, Friedrich Ebert Stiftung and TWARO-ITGLWF). Available at: http://www.fes.or.id/fes/download/Survey_Result_Cambodia.pdf (accessed 2 August 2012).

Cambodia Institute of Development Study (CIDS) (2009b). *Study on Minimum Wage for Cambodia's Garment Industry* (Cambodia Institute of Development Study, Oxfam NOVIB, and Community Legal Education Center).

Cambodia Institute of Development Study (CIDS) (2010). Tracking Study of Cambodian Garment Sector Workers Affected by the Global Economic Crisis. Prepared for ILO, UNDP and Better Factories Cambodia, July.

Cammett, M. (2007). 'Business–Government Relations and Industrial Change: The Politics of Upgrading in Morocco and Tunisia'. *World Development* 35(11), 1889–903.

Carew, A. (2003). 'The Politics of Productivity and the Politics of Anti-Communism'. *Intelligence and National Security* 18, 73–91.

Carr, M. and Chen, M. (2004). *Globalisation, Social Exclusion and Work: With Special Reference to Informal Employment and Gender*. Integration Working Paper No. 20 (Geneva: ILO).

Caspersz, D. (2010). 'From Pollyanna to the Pollyanna Principle. A Response to Michael Burawoy's "From Polanyi to Pollyanna: The False Optimism of Global Labour Studies"' (GLJ 1.2). *Global Labour Journal* 1(3), 393–97.

Cattaneo, O., Gereffi, G. and Staritz, C. (2010). *Global Value Chains in a Post-Crisis World: A Development Perspective* (Washington, DC: World Bank).

Chakrabarty, S. and Grote, U. (2009). 'Child Labor in Carpet Weaving: Impact of Social Labeling in India and Nepal'. *World Development* 37(10), 1683–93.

Chen, M., Sebstad, J. and O'Connel, L. (1999). 'Counting the Invisible Workforce: The Case of Homebased Workers'. *World Development* 27(3), 603–10.

China CSR (2010). Available at: http://www.chinacsr.com/en/ (accessed 20 September 2011).

Clawson, D. (2010). '"False" Optimism: The Key to Historic Breakthroughs? A Response to Michael Burawoy's "From Polanyi to Pollyanna: The False Optimism of Global Labour Studies"' (GLJ 1.2). *Global Labour Journal* 1(3), 398–400.

CLB (2009). *Going It Alone: The Workers' Movement in China (2007–2008)* (Hong Kong: China Labour Bulletin). Available at: http://www.clb.org.hk/en/files/File/research_reports/workers%20movement%2007-08.pdf (accessed 16 September 2013).

CNZFE (Consejo Nacional de Zonas Francas de Exportación) (2005). Informe Estadístico del Sector Zonas Francas de la República Dominicana 2005 (Santo Domingo: CNZFE).

CNZFE (2010). Informe del Sector Textil (Santo Domingo: CNZFE).

Coe, N., Dicken, P. and Hess, M. (2008). 'Global Production Networks: Realizing the Potential'. *Journal of Economic Geography* 8, 271–95.

Coe, N. M. and Hess, M. (2013). 'Global Production Networks, Labour and Development'. *Geoforum* 44, 4–9.
Coe, N. M., and Jordhus-Lier, D. (2010). 'Constrained Agency? Re-Evaluating the Geographies of Labour'. *Progress in Human Geography* 35(2), 211–33.
Cohen, J. and Sabel, C. (2006). 'Extra RempublicamNullaJustitia?' *Philosophy & Public Affairs* 34(2), 147–75.
Collins, J. L. (2003). *Threads: Gender, Labor and Power in the Global Apparel Industry* (Chicago and London: Chicago University Press).
Committee on Monitoring International Labor Standards. (2004). *Monitoring International Labour Standards: Techniques and Sources of Information* (Washington, DC: National Research Council).
Cumbers, A., Nativel, C. and Routledge, P. (2008). 'Labour Agency and Union Positionalities in Global Production Networks'. *Journal of Economic Geography* 8, 369–87.
Datta, A. and Rustagi, P. (2010). *Status of Women in Bihar: Exploring Transformation in Work and Gender Relations* (New Delhi: Institute for Human Development, mimeo).
Degraff, D. S. and Levison, D. (2009). 'Children's Work and Mothers' Work: What is the Connection?' *World Development* 37(9), 1569–87.
Deshingkar, P. (2009). *Extending Labour Inspections to the Informal Sector and Agriculture*. Chronic Poverty Research Centre Working Papers No. 154 (Manchester: University of Manchester).
Deuster P., MacDonald, B. and Zimmerman, D. (2009). *Evaluation of USAID/ Cambodia's Labor and Industrial Productivity Activities* (Washington: USAID).
Diallo, Y., Hagemann, F., Etienne, A., Gurbuzer, Y. and Mehran, F. (2010). *Global Child Labour Developments: Measuring Trends from 2004 to 2008*. ILO International Programme on the Elimination of Child Labour (Geneva: ILO).
Dicken, P. (2003). *Global Shift: transforming the global economy*. 4th ed. (London: Sage Publications).
Dirnbach, E. (2009). *Stolen Wages in the Apparel Industry: International Labor Rights Forum*. Available at: http://laborrightsblog.typepad.com/international_labor_right/2009/10/stolen-wages-and-the-global-apparel-industry.html (accessed 2 March 2010).
Doorey, D. J. (2008). 'Can Factory List Disclosure Improve Labor Practices in the Apparel Industry? A Case Study of Nike and Levi Strauss'. *Comparative Research in Law and Political Economy* 4(1), 1–58.
Drèze, J. and Sen, A. (2002), *India: Development and Participation* (New Delhi: Oxford University Press).
Ear, S. (2011). Growth in the Rice and Garment Sectors. In C. Hughes and K. Un (eds), *Cambodia's Economic Transition* (Copenhagen: Nordic Institute of Asian Studies) pp. 70–94.
Ebert, F. C. and Posthuma, A. (2011). *Labour Provisions in Trade Arrangements: Current Trends and Provisions*. Discussion Paper No. 205. (Geneva: International Institute of Labour Studies) Available at: http://www.ilo.org/inst/publication/discussion-papers/WCMS_192807/lang--en/index.htm (accessed 16 September 2013).
Economic Institute of Cambodia (2007). *Export Diversification and Value Added: Addressing the Impact of the Agreement on Textile and Clothing Expiration on Cambodia* (Phnom Penh: Economic Institute of Cambodia).
Edwards, P. K. and Scullion, H. (1982). *The Social Organizations of Industrial Conflict: Control and Resistance in the Workplace* (Oxford: Blackwell Publishing).

Edwards, R. C. (1979). *Contested Terrain: The Transformation of the Workplace in the Twentieth Century* (New York: Basic Books).

Elliott, K. A. and Freeman, R. B. (2003). *Can Labor Standards Improve under Globalization?* (Washington, DC: Institute for International Economics).

Esbenshade, J. (2004). *Monitoring Sweatshops: Workers, Consumers, and the Global Apparel Industry* (Philadelphia, PA: Temple University Press).

Ethical Trading Initiative (2007). *Purchasing Practices: Case Studies to Address the Impact of Purchasing Practices on Working Conditions*. Report from ETI members meeting 29 November 2007 (London: ETI).

Ethical Trading Initiative (2010). *Management Benchmarks on Purchasing Practices*. Available at: *http://www.ethicaltrade.org/resources/key-eti-resources/eti-management-benchmarks issued 26th February 2010* (accessed 12 November 2010).

European Commission (EC) (1991). *PEPPER I Report – Promotion of Employee Participation in Profits and Enterprise Results in the Member States of the European Community*, COM (96) 697 Final, Brussels.

Evans, P. (2008). 'Is an Alternative Globalization Possible?'. *Politics & Society* 36(2), 271–305.

Evgeniev, E. and Gereffi, G. (2008). 'Textile and Apparel Firms in Turkey and Bulgaria: Exports, Local Upgrading and Dependency'. *Economic Studies* 17(3), 148–79.

Export Processing Zones Authority of Kenya (EPZA) (2009). Annual Performance Report 2009 (Nairobi: EPZA).

Fair Labor Association (2008). *Report on the Best Practices in Apparel Purchasing*, Fair Labor Association Conference. Shanghai, China, 24 June.

Fair Labor Association (2010). *Wages Along The Supply Chain: Developments and Responses*. Stakeholder Forum Report. Hong Kong, 25 October.

Fallon, P. and Tzannatos, Z. (1998), *Child Labour: Issues and Directions for the World Bank* (Washington, DC: IBRD).

Fernandez-Stark, K., Frederick, S. and Gereffi, G. (2011). 'The Apparel Global Value Chain: Economic Upgrading and Workforce Development'. In G. Gereffi, K. Fernandez-Stark and P. Psilos (eds), *Skills for Upgrading: Workforce Development and Global Value Chains in Developing Countries* (Durham, NC: Center on Globalization, Governance & Competitiveness, Duke University) pp. 75–131.

Flanagan, R. J. (2005). *Globalization and Labor Conditions: Working Conditions and Worker Rights in a Global Economy* (New York: Oxford University Press).

Forstater, M. (2010). *Sectoral Coverage of the Global Economic Crisis – Implications of the Global Financial and Economic Crisis on the Textile and Clothing Sector*. Sectoral Activities Programme (Geneva: ILO).

Francois, J., Manchin, M., Norberg, H. and Spinanger, D. (2007). *Impacts of Textiles and Clothing Sectors Liberalization on Prices*. EU Commission Final Report 2007-04-18 for the Directorate-General for Trade (Kiel: The Kiel Institute for the World Economy).

Frank, T. A. (2008). 'Confessions of a Sweatshop Inspector', April 2008. Available at: http://www.washingtonmonthly.com/features/2008/0804.frank.html (accessed 13 August 2013).

Frederick, S. and Gereffi, G. (2011). 'Upgrading and Restructuring in the Global Apparel Value Chain: Why China and Asia Are Outperforming Mexico and Central America'. *International Journal of Technological Learning, Innovation and Development* 4(1/2/3), 67–95.

Frederick, S. and Staritz, C. (2012). Developments in the Global Apparel Industry after the MFA Phaseout. In G. Lopez-Acevedo and R. Robertson (eds) *Sewing Success? Employment and Wage Effects of the end of the Multi-fibre Arrangement (MFA)*. Volume I (Washington, DC: The World Bank).

Gap Inc. (2009). *Combatting Child Labour in the Garment Industry*. Available at: http://www.aktiv-gegen-kinderarbeit.de/files/2012/09/Combatting-Child-Labour-in-the-Garment-Industry.pdf (accessed 20 September 2013).

Gee, K. A. (2010). 'Reducing Child Labour through Conditional Cash Transfer: Evidence from Nicaragua's *Red de Proteccion Social*'. *Development Policy Review* 28, 6.

General Agreement on Tariffs and Trade (GATT) (1989). Trade Policy Review Mechanism: Kingdom of Morocco. Report by the Secretariat (Geneva: GATT Secretariat).

Gereffi, G. (1994). 'The Organization of Buyer-Driven Global Commodity Chains: How US Retailers Shape Overseas Production Networks'. In G. Gereffi and M. Korzeniewicz (eds), *Commodity Chains and Global Capitalism* (Westport, CT: Greenwood Press) pp. 95–122.

Gereffi, G. (1999). 'International Trade and Industrial Upgrading in the Apparel Commodity Chain'. *Journal of International Economics* 48, 37–70.

Gereffi, G. (2005). 'The Global Economy: Organization, Governance and Development'. In N. J. Smelser and R. Swedberg (eds), *Handbook of Economic Sociology* (Princeton: Princeton University Press and Russell Sage Foundation) pp. 160–182.

Gereffi, G. and Bair, J. (2010). *Strengthening Nicaragua's Position in the Textile-apparel Value Chain: Upgrading in the Context of the CAFTA-DR Region*. Study commissioned by USAID/CARANA and the National Free Zones Commission (CNZF/Government of Nicaragua), 20 December Available at: http://www.cggc.duke.edu/pdfs/2010-12-20_Gereffi_Bair_Nicaragua-apparel-report.pdf (accessed 13 August 2013).

Gereffi, G. and Christian, M. (2009). 'The Impacts of Wal-Mart: The Rise and Consequences of the World's Dominant Retailer'. *Annual Review of Sociology* 35(1), 573–91.

Gereffi, G. and Frederick, S. (2010). 'The Global Apparel Value Chain, Trade and the Crisis: Challenges and Opportunities for Developing Countries'. In O. Cattaneo, G. Gereffi and C. Staritz (eds), *Global Value Chains in a Postcrisis World. A Development Perspective* (Washington, DC: World Bank) pp. 157–208.

Gereffi, G., Humphrey, J. and Sturgeon, T. (2005). 'The Governance of Global Value Chains'. *Review of International Political Economy*, 12(1), 78–104.

Gereffi, G. and Mayer, F. W. (2004). *The Demand for Global Governance*. Working Papers Series SAN04-02.

Gereffi, G. and Mayer, F. M. (2006). 'Globalization and the Demand for Governance'. In G. Gereffi, *The New Offshoring of Jobs and Global Development. ILO Social Policy Lectures* (Geneva: International Institute for Labour Studies) pp. 39–58.

Gereffi, G. and Memedovic, O. (2003) *The Global Apparel Value Chain: What Prospects for Upgrading for Developing Countries*. Sectoral Studies Series. (Vienna: United Nations Industrial Development Organization (UNIDO)).

Gereffi, G., Spener, D. and Bair, J. (eds) (2002). *Free Trade and Uneven Development: The North American Apparel Industry after NAFTA* (Philadelphia, PA: Temple University Press).

German Gesellschaft für Technische Zusammenarbeit (GTZ). 'Promotion of Social Environmental, and Production Standards in the Readymade Garment

Industry Project in Bangladesh'. http://www.giz.de/themen/en/20176.htm (accessed September 2013).

Ghose, A. K. (2004). 'The Employment Challenge in India'. *Economic and Political Weekly*, 27 November.

Gibbon, P. and Ponte, S. (2005). *Trading Down: Africa, Value Chains, And The Global Economy* (Philadelphia, PA: Temple University Press).

Gibson-Graham, J. K., Resnick, S. and Wolff, R. (eds) (2000). *Class and Its Others* (Minneapolis, MN: University of Minnesota Press).

Global Labour Strategies (2008). Beyond China's New Labour Contract Law. *Global Labour Strategies*. Available at: http://laborstrategies.blogs.com/global_labor_strategies/2008/01/beyond-chinas-n.html (accessed 20 September 2011).

Global Macro Monitor (2011). 'Trend Reversal in Apparel Prices?' Available at: http://macromon.wordpress.com/2011/02/14/trend-reversal-in-apparel-prices/ (accessed 23 April 2013).

Gomberg, W. (1948). *A Trade Union Analysis of Time Study* (Chicago: Science Research Associates).

Gomberg, W. (1968). 'Union Policy Experimentation in a Volatile Industry'. *Labor History* 9(1), Spring, 69–81.

Sergio Gonzalez, J.E. Austin (2007). *Comprehensive Competitiveness Strategy Report For The Garment Sector: Recommendations For Ongoing SABEQ Support In Jordan's Apparel Sector* (Washington, USA).

Gouvea Abras, A.L. (2012). 'Success and Upgrading after the End of the MFA'. In G. Lopez-Acevedo and R. Robertson (eds) (2012). *Sewing Success? Employment, Wages, and Poverty following the End of the Multi-fibre Arrangement* (Washington, DC: World Bank).

Government of the Kingdom of Morocco (GoM) (2011). *Rapport Economique et Financier* (Rabat: Ministère de l'Economie et des Finances).

Grant, A. (1983). *Against the Clock* (London: Pluto Press).

Gregoratti, C. and Miller, D. (2009). *(Re-)Instating Decent Work through Mature Systems of Industrial Relations: The Case of River Rich Cambodia*. Paper for the Conference Regulating for decent work: Innovative labour regulation in a turbulent world, ILO, Geneva, 8–10 July.

H&M (2006). *Corporate Social Responsibility Report 2005* (Stockholm: H&M).

Hale, A. and Wills, J. (eds) (2005). *Threads of Labour* (London: Blackwell Publishing).

Hanlon, J., Barrientos, and Hulme, D. (2010). *Just Give Money to the Poor: The Development Revolution from the Global South* (Sterling, VA: Kumarian Press).

Harris, R. and Qian, C. L. (2010). 'Export-Market Dynamics and the Probability of Firm Closure: Evidence for the United Kingdom'. *Scottish Journal of Political Economy* 57(2), 145–68.

Harrison, A. E. and Scorse, J. (2010). 'Multinationals and Anti-Sweatshop Activism'. *American Economic Review* 100(1), 247–73.

Hart, G. (2002). *Disabling Globalization* (Pietermaritzburg: University of Natal Press).

Henderson, J., Dicken, P., Hess, M., Coe, N. and W-C. Yeung, H. (2002). 'Global Production Networks and the Analysis of Economic Development'. *Review of International Political Economy* 9(3), 436–64.

Hess, M. and Coe, N. (2006). 'Making Connections: Global Production Networks, Standards and Embeddedness in the Mobile Telecommunications Industry'. *Environment and Planning A* 38, 7, 1205–27.

Holmes, R. (2009). 'Emma Watson launches ethical fashion range with People Tree', *The Guardian*, 17 September. Available at: http://www.guardian.co.uk/lifeandstyle/2009/sep/17/emma-watson-ethical-fashion-people-tree (accessed 20 September 2011).

Hopkins, T. and Wallerstein, I. (1977). 'Patterns of Development of the Modern World-System'. *Review* 1(2), 11–145.

Hornbeck, J. F. (2010). *The Haitian Economy and the Hope Act.* Congressional Research Service Report (Collingdale, PA: Diane Publishing).

Hughes, A. (2006). 'Learning to Trade Ethically: Knowledgeable Capitalism, Retailers and Contested Commodity Chains'. *Geoforum* 37(6), 1008–20.

Hughes, C. (2007). 'Transnational Networks, International Organizations and Political Participation in Cambodia: Human Rights, Labour Rights and Common Rights'. *Democratization* 14(5), 834–52.

Hughes, C. and Un, K. (2011). 'Cambodia's Economic Transformation: Historical and Theoretical Frameworks'. In C. Hughes and K. Un (eds), *Cambodia's Economic Transition* (Copenhagen: Nordic Institute of Asian Studies) pp. 1–26.

Hughes, C., Wrigley, N. and Buttle, M. (2008). 'Global Production Networks, Ethical Campaigning, and the Embeddedness of Responsible Governance'. *Journal of Economic Geography* 8(3), 345–67.

Humphrey, J. (2004). *Upgrading in Global Value Chains.* Integration Working Paper No. 28 (Geneva: ILO).

Humphrey, J. and Schmitz, H. (2002). 'How Does Insertion in Global Value Chains Affect Upgrading in Industrial Clusters?' *Regional Studies* 36, 9.

Hussain-Khaliq, S. (2004). 'Eliminating Child Labour from the Sialkot Soccer Ball Industry – Two Industry Led Approaches'. *Journal of Corporate Citizenship*, 13.

Institute of Applied Manpower Research (IAMR) and Planning Commission (2011). *India Human Development Report* (New Delhi: Oxford University Press).

Ichniowski, C., Shaw, K. and Prennushi, B. (1997). 'The Effects of Human Resource Management Practices on Productivity: A Study of Steel Finishing Lines'. *American Economic Review* 87(3), 291–313.

ILO (International Labour Office) (1992). *International Labour Conventions and Recommendations 1919–1991*, 2 volumes (Geneva: ILO).

ILO (International Labour Office) (1999). *Decent Work. Report of the Director-General to the 89th Session of the International Labour Conference* (Geneva: ILO).

ILO (International Labour Office) (2000). *L'industrie du textile-habillement au Maroc: les besoins des chefs d'entreprise et les conditions des femmes dans les PME.* Sectoral Activities Working Paper No. 159 (Geneva: ILO).

ILO (International Labour Office) (2008a). *Global Wage Report 2008/09 – Minimum Wages and Collective Bargaining towards Policy Coherence* (Geneva: ILO).

ILO (International Labour Office) (2008b). ILO Declaration on Social Justice for a Fair Globalization, adopted by the International Labour Conference at its 97th Session, Geneva, 10 June.

ILO (International Labour Office) (2010). *Global Wage Report 2010/11: Wage Policies in Times of Crisis* (Geneva: ILO).

ILO (International Labour Office) (2010). *Better Factories Cambodia.* Available at: http://www.betterfactories.org/ilo/ (accessed 20 September 2011).

Impactt/Traidcraft (2009). *Material Concerns* (London, Impactt/Traidcraft).

International Monetary Fund (IMF) (2010). *Regional Economic Outlook: Asia and Pacific – Consolidating the Recovery and Building Sustainable Growth*, October, World Economic and Financial Surveys (Washington, DC: IMF).

International Monetary Fund (IMF), ILO, Organisation for Economic Co-operation and Development (OECD), Statistical Office of the European Communities (Eurostat), United Nations Economic Commission for Europe and World Bank (2009). *Export and Import Price Index Manual. Theory and Practice* (Washington, DC: IMF).

International Textile Garment and Leather Workers Federation (2009). *Programme of Action* (Brussels: ITGLWF).

International Trade Union Confederation, IndustriALL, Clean Clothes Campaign, Uni (2012) *The UN Guiding Principles on Business and Human Rights and the Human Rights of Workers to Form or Join Trade Unions and to Bargain Collectively*, Brussels.

International Textile, Garment and Leather Workers' Federation (ITGLWF) (2006). 'Making the "Social" in CSR Sustainable'.

ITUC (2008). *Human Trafficking for Labour Exploitation in Romania*. ITUC Submission to UPR in Romania, February 2008, Brussels.

Iwanow, H., McEachern, M. and Jeffrey, A. (2005). 'The Influence of Ethical Trading Policies on Consumer Apparel Purchase Decisions: A Focus on GAP Inc'. *International Journal of Retail and Distribution Management* 33(5), 371–87.

Jassin-O'Rourke Group (2002). *Global Competitiveness Report: Selling to Full Package Providers* (New York: Jassin-O'Rourke Group, LLC).

Jassin-O'Rourke Group (2008). *Global Apparel Manufacturing Labor Cost Analysis 2008* (New York: Jassin-O'Rourke Group, LLC).

Jeffrey, M. and Evans, N. (2011). *Costing for the Fashion Industry* (London: Berg).

Jenkins, R. (2001). *Corporate Codes of Conduct: Self-regulation in a Global Economy* (Geneva: United Nations Research Institute for Social Development).

Jenkins, R. (2002). 'Political Economy of Codes of Conduct'. In R. Jenkins, R. Pearson, and G. Seyfang (eds), *Corporate Responsibility and Labour Rights: Codes of Conduct in the Global Economy* (New York: Routledge) pp. 13–30.

Jenkins, J. (2010). *Courage, Knowledge and the Power of Organization*. Paper to the Annual International Labour Process Conference, 15–17 March, Rutgers University, New Jersey.

Jenkins, R., Pearson, R. and Seyfang, G. (eds) (2002). *Corporate Responsibility and Labour Rights*. (London: Earthscan Publications).

Just-style (2010). 'Cambodia: Talks to resolve pay dispute make slow progress'. Available at: http://www.just-style.com/news/talks-to-resolve-pay-dispute-make-slow-progress_id109061.aspx (accessed 24 April 2013)

Kanawaty, G. (ed.) (1992). *Introduction to Work Study* (Geneva: ILO).

Kanbur, R. (2009). 'Conceptualising Informality: Regulation and Enforcement'. *Indian Journal of Labour Economics*, 52, 1.

Kang, K., Sok, S. and Dannet, L. (2009). *Rapid Assessment of the Impact of the Financial Crisis in Cambodia*. (Bangkok: ILO Regional Office for Asia and the Pacific).

Kaplinsky, R. (2000). 'Globalisation and Unequalisation: What Can be Learned from Value Chain Analysis'. *The Journal of Development Studies* 37(2), 117.

Kaplinsky, R. and Readman, J. (2005). 'Globalization and Upgrading: What Can (and Cannot) be Learnt from International Trade Statistics in the Wood Furniture Sector?' *Industrial and Corporate Change* 14(4), 679–703.

Keohane, R. O. (2002). *Power and Governance in a Partially Globalized World* (New York and London: Routledge).

Khanam, R. (2006). 'Child Labour in Bangladesh: Trends, Patterns and Policy Options'. *Asian Profile* 34(6), 593–608. Available at: http://mpra.ub.uni-muenchen.de/8008/1/MPRA_paper_8008.pdf (accessed 13 August 2013).

Kline, J. M. (2010). *Alta Gracia: Branding Decent Work Conditions. Will College Loyalty Embrace 'Living Wage' Sweatshirts?* (Washington, DC: Georgetown University).

Knorringa, P. and Pegler, L. (2006). 'Globalisation, Firm Upgrading and Impacts on Labour'. *Tijdschrift voor economische en sociale geografie* 97(5), 470–79.

Laleva, I. and Musiolek, B. (1999). *Conditions in the Savina Factory in Sandanski*, November 1999. Clean Clothes Campaign. Available at: http://archive.cleanclothes.org/campaigns-list/814-clean-clothes-campaigns-set-to-score-at-euro-2000 (accessed 20 September 2013).

Lambert, R. (2010). 'Unionism in One Country Is No Longer an Option. A Response to Michael Burawoy's "From Polanyi to Pollyanna: The False Optimism of Global Labour Studies"' (GLJ 1.2). *Global Labour Journal* 1(3), 388–92.

Lamming, R., Caldwell, N., Phillips, W. and Harrison, D. (2005). 'Sharing Sensitive Information in Supply Relationships: The Flaws in One-way Open-book Negotiation and the Need for Transparency'. *European Management Journal* 23(5), 554–63.

Lan, T. and Pickles, J. (2011). *China's New Labour Contract Law: State Regulation and Worker Rights in Global Production Networks*. Capturing the Gains Working Paper No. 5. Available at: http://www.capturingthegains.org/pdf/ctg-wp-2011-05.pdf (accessed 13 August 2013).

Lazaro, F. (2012). *Cambodian Garment Worker Justice*. Public Broadcasting Service. Available at: http://www.pbs.org/wnet/religionandethics/episodes/may-18-2012/cambodia-garment-worker-justice/11033/ (accessed 2 August 2012).

Lee, C. H. (2006). 'Recent Industrial Relations Developments in China and Viet Nam: The Transformation of Industrial Relations in East Asian Transition Economies'. *Journal of Industrial Relations* 48(3), 415–29.

Lee, C. K. (2007). *Against the Law: Labour Protests in China's Rustbelt and Sunbelt* (Berkeley: University of California Press).

Lee, C. K. (2008). 'Raw Encounters: Chinese Managers, African Workers, and the Politics of Casualization in Africa's Chinese Enclaves'. *The China Quarterly* 199, 647–66.

Leiten, G. K. (2002). 'Child Labour in India: Disentangling Essence and Solutions'. *Economic and Political Weekly*, 28 December.

Leiten, G. K. and Rustagi, P. (2006). *The Nowhere Children: A Study of Daily Practice and Hidden Concerns*, Report submitted to IDPAD.

Lesotho Bureau of Statistics (2010). *Statistical Yearbook 2010* (Maseru: Bureau of Statistics).

Lezama, M., Webber, B. and Dagher, C. (2004). *Sourcing Practices in the Apparel Industry: Implications for Garment Exporters in Commonwealth Developing Countries* (London: Commonwealth Secretariat Special Advisory Services Division).

Li, M. and Edwards, P. (2008). 'Work and Pay in Small Chinese Clothing Firms: A Constrained Negotiated Order'. *Industrial Relations Journal* 39(4), 296–313.

Licadho (2012). *Briefing Paper: Cambodia's Draft Law on the Management and Use of Agricultural Land* (Phnom Penh: Licadho). Available at: http://www.

licadho-cambodia.org/reports/files/169LICADHOBriefingDraftAgricultural Law-English.pdf (accessed 2 August 2012).
Locke, R. (2013 forthcoming). *Promoting Labour Rights in a Global Economy* (New York: Cambridge University Press).
Locke, R. M., Rissing, B. A. and Pal, T. (2012). 'Complements or Substitutes? Private Codes, State Regulation and the Enforcement of Labour Standards in Global Supply Chains'. *British Journal of Industrial Relations* 50.
Locke, R. and Romis, M. (2007). 'Improving Work Conditions in a Global Supply Chain'. *MIT Sloan Management Review* 48(2), 54–62.
Locke, R. M. and Romis, M. (2010). 'The Promise and Perils of Private Voluntary Regulation: Labor Standards and Work Organization in Two Mexican Garment Factories'. *Review of International Political Economy* 17(1), 45–75.
Locke, R. M., Qin, F. and Brause, A. (2007). 'Does Monitoring Improve Labor Standards: Lessons from Nike'. *Industrial and Labor Relations Review* 61(1), 3–31.
Locke, R., Amengual, A. and Mangla, A. (2009). *Virtue out of Necessity, Compliance, Commitment and the Improvement of Labour Conditions in Global Supply Chains.* MIT Sloan Working Paper No. 4719–08.
Locke, R., Kochan, K., Romis, M. and Qin, F. (2007). 'Beyond Corporate Codes of Conduct: Work Organization and Labour Standards at Nike's Suppliers'. *International Labour Review* 146(1–2), 21–39.
Lopez-Acevedo, G. and Robertson, R. (eds) (2012). *Sewing Success? Employment, Wages, and Poverty following the End of the Multi-fibre Arrangement* (Washington, DC: World Bank).
Lukas, K., Plank, L. and Staritz, C. (2010). *Securing Labour Rights in Global Production Networks: Legal Instruments and Policy Options* (Vienna: Chamber of Labour).
Malam, L. (2012). *Sewing the Line: A Qualitative Baseline Analysis of the Risks and Opportunities Posed for Young Women By Migration From Rural Lao to Vientiane For the Purpose of Employment in the Garment Manufacturing Industry* (Vientiane, People's Democratic Republic of Lao: CARE International).
Mamic, I. (2003). *Business and Code of Conduct Implementation* (Geneva: ILO).
Maquila Solidarity Network (2008). *Clearing the Hurdles* (Toronto: MSN).
Marks & Spencer (2010). *Marks & Spencer Launches Programme to be World's Most Sustainable Retailer by 2015.* Press release 1 March. Available at: http://corporate.marksandspencer.com/investors/press_releases/sustainableretailer (accessed 17 September 2013).
Mason, R. and Weeds, H. (2010). 'Investment, Uncertainty and Pre-emption'. *International Journal of Industrial Organization* 28(3), 278–87.
Mayer, F. W. (1998). *Interpreting NAFTA: The Science and Art of Political Analysis* (New York: Columbia University Press).
Mayer, F. and Pickles, J. (2010). *Re-Embedding Governance: Global Apparel Value Chains and Decent Work,* Capturing the Gains Working Paper 2010/01.
Mayer, F. and Gereffi, G. (2010). 'Regulation and Economic Globalization: Prospects and Limits of Private Governance'. *Business and Politics* 12(3), Article 11.
Mayer, F. and Posthuma, A. (2012). *Governance in a Value Chain World.* Capturing the Gains briefing note. Available at: http://www.capturingthegains.org/pdf/CTG-Governance.pdf (accessed 17 September 2013).
McGrath-Champ, S., Herod, A. and Rainnie, A. (eds) (2010). *Handbook of Employment and Society: Working Space* (Cheltenham: Edward Elgar).

Mehta, B. and Sherry, K. (2009). 'Wages and Productivity of Child Labour: A Case of the Zardosi Industry'. *Indian Journal of Labour Economics* 52(4), 651–70.

Merk, J. (2009). *Stitching A Decent Wage Across Borders: The Asia Floor Wage Proposal* (New Delhi: Asia Floor Wage Campaign).

Milberg, W. and Winkler, D. (2011). 'Economic and Social Upgrading in Global Production Networks: Problems of Theory and Measurement'. *International Labour Review* 150(3–4), 341–65.

Miller, D. (2005). 'Transition in the Post-apartheid Regional Workplace'. In E. Webster and K. Von Holdt (eds), *Beyond the Apartheid Workplace: Studies in Transition* (Scottsville: University of KwaZulu-Natal Press) pp. 243–65.

Miller, D. (2010). 'Towards Sustainable Labour Costing in the Global Apparel Industry: Some Evidence from UK Fashion Retail'. *Proceedings of the Centenary Conference of the Textiles Institute*, 3–4 November 2010.

Miller, D. (2011). 'Global Social Relations and Corporate Social Responsibility in Outsourced Apparel Supply Chains: The Inditex Global Framework Agreement'. In K. Papadakis (ed.), *Shaping Global Industrial Relations: The Impact of International Framework Agreements*. (Basingstoke and Geneva: Palgrave Macmillan and ILO), 179–200.

Miller, D. and Williams, P. (2008). 'What Price a Living Wage? Implementation Issues in the Quest for Decent Wages in the Global Apparel Sector'. *Global Social Policy* 9(1), 99–125.

Miller, D., Nuon, V., Aprill, C. and Certeza, R. (2009). 'Business as Usual? Governing the Supply Chain in Clothing – Post MFA phase-out: The case of Cambodia'. *International Journal of Labor Research* 1(1), 10–33.

Miller, D., Turner, S. and Grinter, T. (2011). 'Back to the Future? Mature Systems of Industrial Relations Approaches and Governance Issues in Outsourced Apparel Supply Chains'. Unpublished paper.

Morris, M. and Barnes, J. (2009). *Globalization, the Changed Global Dynamics of the Clothing and Textile Value Chains and the Impact on Sub-Saharan Africa*. UNIDO Research and Statistics Branch Working Paper No. 10 (Vienna: UNIDO).

Morris, M., Staritz, C. and Barnes, J. (2011). 'Value Chain Dynamics, Local Embeddedness, and Upgrading in the Clothing Sectors of Lesotho and Swaziland'. *International Journal of Technological Learning, Innovation and Development* 4(1/2/3), 96–119.

Mosse, D. (2010). 'A Relational Approach to Durable Poverty, Inequality and Power'. *Journal of Development Studies* 47(7), 1156–78.

Maquila Solidarity Network (MSN), Honduran Independent Monitoring Team (EMIH), and Professionals for Corporate Social Auditing (PASE) (2012). *Can National Competitiveness Strategies Include Decent Work? A comparative study of the garment export industries in Nicaragua and Honduras*, June.

Mu, S. and Wikstrom, C. (2012). 'Land Grabs in Cambodia'. *The New York Times* 18 July.

Murray, C. (1981). *Families Divided: The Impact of Migrant Labour in Lesotho* (New York: Cambridge University Press).

Murray, I. (2008). 'WWB Gender Study: The Capacity of Poor Women to Grow Their Business in the Dominican Republic'. *Global Urban Development* 4(2), November.

Nadvi, K. (2008). 'Global Standards, Global Governance and the Organization of Global Value Chains'. *Journal of Economic Geography* 8(3), 323–43.

Nadvi, K. and Wältring, F. (2001). *Making Sense of Global Standards*. Draft IDS-INEP Working Paper (Brighton, UK: Institute for Development Studies).

Nashir, U., Mohammad, M. H. and Gunter, B. G. (2009). *Physical and Psychological Implications of Risky Child Labor: A Study in Sylhet City, Bangladesh*. Bangladesh Development Research Working Paper Series, BDRWPS No. 8 (Falls Church, VA: Bangladesh Development Research Center (BDRC)) July.

Nathan Associates (2005). *Measuring Competitiveness and Productivity in Cambodia's Apparel Industry* (Washington, DC: USAID).

Nathan Associates and Werner International (2007). *Factory-Level Value Chain Analysis of Cambodia's Apparel Industry* (Washington, DC: USAID).

Nathan, D., Kumar, A. and Sapra, S. (2013). *Labour Conditions in GPNs: Garment Manufacture*. Paper prepared for the IHD-ICSSR research programme on 'Globalization and Labour', mimeo, New Delhi.

National Statistics Centre (2006). *Addressing the Impacts of the Phasing Out of Textiles and Clothing Quotas in Lao PDR*. (Vientiane: National Statistics Centre).

Nicholls, A. and Opal, C. (2005). *Fairtrade: Market-Driven Ethical Consumption* (London: Sage).

Nichols, S. (2001). 'All Eyes on Americas at MAGIC's Sourcing Summit'. *Apparel* 31 August. Available online at: http://apparel.edgl.com/case-studies/all-eyes-on-americas-at-magic-s-sourcing-summit75261?referaltype=newsletter (accessed 2 October 2011).

Nova, S. (2013). 'How Apparel Industry Pricing Practices Undermine Labour Rights Progress'. In J. Bair, M. Dickson and D. Miller (eds) *Social Labelling and the Global Fashion Industry* (London: Routledge).

Oka, C. (2010a). 'Accounting for the Gaps in Labour Standard Compliance: The Role of Reputation-Conscious Buyers in the Cambodian Garment Industry'. *European Journal of Development Research* 22(1), 59–78.

Oka, C. (2010b). 'Channels of Buyer Influence and Labor Standard Compliance: The Case of Cambodia's Garment Sector'. *Advances in Industrial and Labor Relations* 17, 153–83.

Oka, C. (2012). *Does Better Labour Standard Compliance Pay? Linking Labour Standard Compliance and Supplier Competitiveness*. Better Work discussion paper No. 5 (Geneva: ILO).

O'Rourke, D. (1997). *Smoke From A Hired Gun: A Critique of Nike's Labor and Environmental Auditing in Vietnam as Performed by Ernst and Young* (San Francisco, CA: Transnational Resource and Action Center).

O'Rourke, D. (2002). 'Monitoring the Monitors: A Critique of Corporate Third-Party Labour Monitoring'. In R. Jenkins et al. (eds) *Corporate Responsibility and Labour Rights* (London: Earthscan Publications).

O'Rourke, D. (2011). 'Citizen Consumer'. *Boston Review*, November/December. Available at: http://www.bostonreview.net/novemberdecember-2011 (accessed 13 August 2013).

Oxfam Australia (2011). 'Application of Australian Competition Law to Campaign to Improve the Wages and Conditions of Certain Workers'. Memorandum from DLA Phillips Fox, 8 March 2011.

Oxfam International (2004). *Trading Away Our Rights: Women Working in Global Supply Chains* (Oxford: Oxfam International).

Palacios, J. J. (2001). *Globalization's Double Movement: Societal Responses to Market Expansion in the 21st Century*. Proceedings of the Eighth Karl Polanyi International Conference 'Economy and Democracy', Universidad Nacional Autonoma de Mexico, Mexico City, 14–16 November.

Palpacuer, F. (2006). 'Globalization and Corporate Governance: Issues for Management Researchers'. *Society and Business Review* 1(1).

Palpacuer, F. (2008). 'Bringing the Social Context Back in: Governance and Wealth Distribution in Global Commodity Chains'. *Economy and Society* 37(3), 393–419.

Papadakis, K. (ed.) (2008). *Cross-Broader Social Dialogue and Agreements* (Geneva: International Institute for Labour Studies/ILO).

Pellegrin, J. (2001). *The Political Economy of Competitiveness in an Enlarged Europe* (Basingstoke/New York: Palgrave).

Pfeiffer, T. (2007). 'Zara Owner Lays Down Ethics for Moroccan Suppliers'. *Reuters*, 27 June.

Phillips, N. (2013). 'Unfree Labour and Adverse Incorporation in the Global Economy: Comparative Perspectives from Brazil and India'. *Economy and Society* 42(2), 171–96

Phillips, N. and Sakamoto, L. (2012). 'Global Production Networks, Chronic Poverty and 'Slave Labor' in Brazil'. *Studies in Comparative International Development* 47(3), 287–315.

Phillips, N., Bhaskaran, R., Nathan, D. and Upendranadh, C. (2011). *Child Labour in Global Production Networks: Poverty, Vulnerability and 'Adverse Incorporation' in the Delhi Garments Sector*. Chronic Poverty Research Centre Working Papers series, No. 177, June (Manchester and New Delhi: Chronic Poverty Research Centre).

Phouxay, K. and Tollefsen, A. (2011). 'Rural-Urban Migration, Economic Transition, and Status of Female Industrial Workers in Lao PDR'. *Population, Space and Place* 17(5), 421–34.

Pickles, J. (2012a). 'Death on the Line'. *Occasional links and commentary on economics, culture and society*. Available at: http://anticap.wordpress.com/2012/12/14/death-on-the-line/ (accessed 13 August 2013).

Pickles, J. (2012b). *Economic and Social Upgrading in Apparel Global Value Chains: Public Governance and Trade Policy*. Capturing the Gains Working Paper No. 13. Available at: http://www.capturingthegains.org/publications/workingpapers/wp_201213.htm (accessed 13 August 2013).

Pickles, J. and Smith, A. (2010). 'Clothing Workers after the Worker States: The Consequences for Work and Labour of Outsourcing, Nearshoring and Delocalization in Postsocialist Europe'. In S. McGrath-Champ et al. (eds) *Handbook of Employment and Society* (Cheltenham: Edward Elgar) pp. 106–23.

Pickles, J., Smith, A., Roukova, P., Begg, R. and Bucek, M. (2006). 'Upgrading and Diversification in the East European Apparel Industry: Competitive Pressure and Production Networks in the Clothing Industry'. *Environment and Planning A* 38(12), 2305–24

Pigman, G. A. (2005). *Civilizing World Trade*. Paper presented at the annual meeting of the International Studies Association, Hilton Hawaiian Village, Honolulu, Hawaii.

Pincheson, E. (1995). *East European Textiles and Clothing: An Industry in Transition* (London: Pearson Professional).

Piore, M. J. and Shrank, A. (2006). 'Trading Up: An Embryonic Model for Easing the Costs of Free Markets'. *Boston Review* 31(5), 11–14.

Piore, M. J. (2008). *Second Thoughts: On Economics, Sociology, Neoliberalism, Polanyi's Double Movement and Intellectual Vacuums*. MIT Industrial Performance Centre Working Paper Series Nos. 08–004, 22 July (Cambridge, MA: MIT).

Plank, L. and Staritz, C. (2009). 'Global Production Networks, Uneven Development, and Workers: Experiences from the Romanian Apparel Sector'. *Austrian Journal for Development Studies* 25(2), 62–87.

Plank, L. and Staritz, C. (2011). *Working in Global Production Networks: Industrial and Social Upgrading Processes in Apparel and Electronics in Central and Eastern Europe*. Dissertation, Graz and Vienna: University of Graz/Vienna University of Economics and Business.

Plankey-Videla, N. (2012). *We Are in this Dance Together: Gender, Power, and Globalization at a Mexican Garment Firm* (New Brunswick, NJ: Rutgers University Press).

Polanyi, K. (2001). *The Great Transformation: The Political and Economic Origins of Our Time (1944)* (New York: Beacon Press).

Polaski, S. (2006). 'Combining Global and Local Forces: the Case of Labor Rights in Cambodia'. *World Development* 34(5), 919–32.

Polaski, S. (2009). *Harnessing Global Forces to Create Decent Work in Cambodia*. International Institute of Labour Studies, Research Series No. 119 (Geneva: International Institute of Labour Studies).

Polaski, S. (2004). *Cambodia Blazes a New Path to Economic Growth and Job Creation*. Carnegie Paper No. 51.

Pollin, R., Burns, J. and Heintz, J. (2004). 'Global Apparel Production and Sweatshop Labour: Can Raising Retail Prices Finance Living Wages?' *Cambridge Journal of Economics* 28(2), 153–71.

Pongsawat, P. (2007). *Border Partial Citizenship, Border Towns, and Thai-Myanmar Cross-Border Development: Case Studies at the Thai Border Towns*. PhD Dissertation City and Regional Planning, University of California, Berkeley.

Posthuma, A. (2010). 'Beyond "Regulatory Enclaves": Challenges and Opportunities to Promote Decent Work in Global Production Networks'. In A. Posthuma and D. Nathan (eds), *Labour in Global Production Networks in India* (Oxford: Oxford University Press), pp. 57–80.

Posthuma, A. and Nathan, D. (eds) (2010). *Labour in Global Production Networks in India* (Oxford: Oxford University Press).

ProNicaragua (2011). *La Industria de Textiles y Confección en Nicaragua*. August (Managua: ProNicaragua).

Rainnie, A., Herod, A. and McGrath-Champ, S. (2011). 'Review and Positions: Global Production Networks and Labour'. *Competition and Change* 15(2), 155–69

Raworth, K. and Kidder, T. (2009). 'Mimicking "Lean" in Global Value Chains – It's the Workers Who Get Leaned On'. In J. Bair (ed.), *Frontiers of Commodity Chain Research* (Stanford: Stanford University Press).

Rivoli, P. (2005) *The Travels of a T-Shirt in the Global Economy* (New York: John Wiley & Sons Inc.).

Roberts, A. (2010). 'H&M, Zara Fast Fashion Pressures Luxury Labels to Speed Up'. Bloomberg, 30 September. Available at: http://www.bloomberg.com/news/2010-09-30/h-m-zara-set-pace-as-fast-fashion-mentality-spreads-to-luxury-companies.html (accessed 13 August 2013).

Roberts, S. (2003). 'Supply Chain Specific? Understanding the Patchy Success of Ethical Sourcing Initiatives'. *Journal of Business Ethics* 44(2–3), 159–70.

Robertson, R. (2011). *Apparel Wages Before and After Better Factories Cambodia.* Better Work Discussion Paper No. 3 (Geneva: ILO and IFC).

Robertson, R., Dehejia, R., Brown, D. and Ang, D. (2011). *Labor Law Compliance and Human Resource Management Innovation: Better Factories Cambodia.* Better Work Discussion Paper No. 1 (Geneva: ILO and IFC).

Rodriguez-Garavito, C. (2005). 'Global Governance and Labor Rights: Codes of Conduct and Anti-Sweatshop Struggles in Global Apparel Factories in Mexico and Guatemala'. *Politics and Society* 33(2), 203–33.

Rogers, T. (2012). *Nicaragua Sets Pace for Region under CAFTA.* Ticotimes, 7 May. Available at: http://www.ticotimes.net/Region/Nicaragua-sets-pace-for-region-under-CAFTA_Friday-May-11-2012/ (accessed 14 August 2013).

Rossi, A. (2011). *Economic and Social Upgrading in Global Production Networks: the Case of the Garment Industry in Morocco.* DPhil dissertation, Institute of Development Studies, University of Sussex.

Rossi, A. (2013). 'Does Economic Upgrading Lead to Social Upgrading in Global Production Networks? Evidence from Morocco'. *World Development* 46, 223–33.

Rossi, A. (2013). 'The Impact of the Fibre Citoyenne Label on the Moroccan Garment Industry and its Workers'. In J. L. Bair., M. Dickson and D. Miller (eds) *Workers' Rights and Labor Compliance in Global Supply Chains: Is a Social Label the Answer?* (London: Routledge).

Rossi, A. and Robertson, R. (2011). 'Better Factories Cambodia – An Instrument for Improving Industrial Relations in a Transnational Context'. In K. Papadakis (ed.), *Shaping Global Industrial Relations* (Basingstoke and Geneva: Palgrave Macmillan and ILO) pp. 220–42.

Ruggie, J. G. (1982). 'International Regimes, Transactions, and Change: Embedded Liberalism in the Postwar Economic Order'. *International Organization* 36(2), 379–415.

Ruwanpura, K. and Wrigley, N. (2010). 'The Costs of Compliance? Views of Sri Lankan Apparel Manufacturers in Times of Global Economic Crisis'. *Journal of Economic Geography* 11(6), 1–19.

Schrank, A. (2004). 'Ready-to-Wear Development? Foreign Investment, Technology Transfer, and Learning by Watching in the Apparel Trade'. *Social Forces* 83(1), 123–56.

Seidman, G. (2009). *Social Labeling in Export Supply Chains: Can Voluntary Certification Programs End Child Labour?* Centre for the Advanced Study of India, University of Pennsylvania, 21 December. Available at: http://casi.ssc.upenn.edu/iit/seidman (accessed 20 September 2011).

Selwyn, B. (2012). 'Beyond Firm-Centrism: Re-Integrating Labour and Capitalism into Global Commodity Chain Analysis'. *Journal of Economic Geography* 12, 205–26.

Seo, J. W. (2011). *Excessive overtime, workers, and productivity.* Better Work Discussion Paper No. 2 (Geneva: ILO and IFC).

Sharma, A. N., Sharma, R. and Raj, N. (2004). *Child Labour in Carpet Industry* (New Delhi: Institute for Human Development).

Shea, A., Nakayama, M. and Heymann, J. (2010). 'Improving Labor Standards in Clothing Factories'. *Global Social Policy* 10(1), 85–110.

Siegmann, K. A. (2008). 'Soccer Ball Production for Nike in Pakistan'. *Economic and Political Weekly*, May.

Silici, L. (2010). 'Conservation Agriculture and Sustainable Crop Intensification in Lesotho'. *Integrated Crop Management* 10.

Smith, A., Pickles J., Bucek, M., Begg, R. and Roukova, P. (2008). 'Reconfiguring "Post-Socialist" Regions: Cross-Border Networks and Regional Competition in the Slovak and Ukrainian Clothing Industry'. *Global Networks* 8(3), 281–301.

Solidarity Center. (2011). *A Post-Earthquake Living Wage Estimate for apparel Workers in the Sonapi Export Processing Zone, Port au Prince*. (Washington, DC: Solidarity Center).

SOMO Centre for Research on Multinational Corporations (2003). *Pricing in the Global Apparel Industry* (Amsterdam: SOMO).

Sontag, D. (2012). 'Earthquake Relief Where Haiti Wasn't Broken'. *The New York Times*, 6 July.

Springer, S. (2008). 'Violence, Democracy, and the Neoliberal "Order": The Contestation of Public Space in Posttransitional Cambodia'. *Annals of the Association of American Geographers* 99(1), 138–62.

Standing, G. (1999). 'Global Feminization through Flexible Labour: A Theme Revisited'. *World Development* 27(3), 583–602.

Stanford Law School and Workers Rights Consortium (2013). *Monitoring in the Dark: An Evaluation of the International Labour Organization's Better Factories Cambodia Monitoring and Reporting Program* (Stanford, CA: International Human Rights and Conflict Resolution Clinic, Mills Legal Clinic, Stanford Law School and Workers Rights Consortium).

Staritz, C. and Frederick, S. (2012). 'Summaries of the Country Case Studies on Apparel Industry Development, Structure and Policies'. In G. Lopez-Acevedo and R. Robertson (eds), *Sewing Success. Employment and Wage Effects of the End of the Multi-fibre Arrangement (MFA)*, volume II (Washington, DC: World Bank).

Staritz, C. (2011). *Making the Cut? Low-Income Countries and the Global Clothing Value Chain in a Post-Crisis World* (Washington, DC: World Bank).

Stewart, M. and Hunter, L. (1964). *The Needle is Threaded: The History of an Industry* (London: Heinemann).

Tesch, R. (1990). *Qualitative Research: Analysis Types and Software Tools* (Philadelphia, PA: Routledge Falmer).

Tewari, M. (2006). 'Is Price and Cost Competitiveness Enough for Apparel Firms to Gain Market Share in the World after Quotas? A Review'. *Global Economy Journal* 6(4), 1–46.

Tewari, M. (2008). 'Varieties of Global Integration: Navigating Institutional Legacies and Global Networks in India's Garment Industry'. *Competition & Change* 12(1), 49–67.

Textile Outlook International (2009). *Report Summary: Trends in World Textile and Clothing Trade*, Issue 143, February.

Textiles Intelligence (1997). *Internationalisation of European Textiles and Clothing Production: Eastern Europe and North Africa*. Special Report No. 2643 (Brussels and London: Textiles Intelligence/OETH).

The Economist (2005). 'Business: The Future of Fast Fashion; Inditex'. Issue 63, 18 June.

The Economist (2008). 'The Spanish Legion', 6 November.

Thoburn, J. (2009). *The Impact of World Recession on the Textile and Garment Industries in Asia*. UNIDO Research and Statistics Branch Working Paper No. 17 (Vienna: UNIDO).

Tickell, A. and Peck, J. (2003). 'Making Global Rules: Globalisation or Neoliberalisation'. In J. Peck and H. Yeung (eds), *Remaking the Global Economy* (London: Sage Publications) pp. 163–82.

Tilly, C. (1998). *Durable Inequality* (Berkeley and Los Angeles, CA: University of California Press).

Tokatli, N. (2007). 'Asymmetrical Power Relations and Upgrading among Suppliers of Global Clothing Brands: Hugo Boss in Turkey'. *Journal of Economic Geography* 7(1), 67–92.

Tokatli, N. (2008). 'Global Sourcing: Insights from the Global Clothing Industry – The Case of Zara, a Fast Fashion Retailer'. *Journal of Economic Geography* 8(1), 21–38.

Turner, M. (2007). 'Society Must Be Protected: Polanyi's "Double Movement" and the Regulation of Conflict Goods'. *Journal of Corporate Citizenship* special issue on 'Karl Polanyi and the Corporate Contribution to Global Peace and Security', Summer, 85–99.

United Nations Industrial Development Organization (UNIDO) (2011). *Industrial Value Chain Diagnostics: An Integrated Tool* (Vienna: UNIDO).

Vaughan-Whitehead, D. (1995). *Workers' Financial Participation – East-West Experiences*, Labour–Management Relations Series (Geneva: ILO).

Vaughan-Whitehead, D. (2010). *Fair Wages: Strengthening Corporate Social Responsibility* (London: Edward Elgar).

Von Holdt, K. and Webster, E. (2005). 'Workplace Restructuring and the Crisis of Social Reproduction'. In E. Webster and K. Von Holdt (eds), *Beyond the Apartheid Workplace: Studies in Transition* (Scottsville: University of KwaZulu-Natal Press) pp. 3–40.

Vorley, B., Fearne, A. and Ray, D. (2007). *Regoverning Markets: A Place for Small-scale Producers in Agrifood Chains* (Farnham, UK: Ashgate).

Wang, H., Appelbaum, R., Degiuli, F. and Lichtenstein, N. (2009). 'China's New Labour Contract Law: Is China Moving Towards Increased Power for Workers?' *Third World Quarterly* 30(3), 485–501.

Waterman, P. (2011). 'Beyond Polanyi and Pollyanna – Oscar Wilde?' *Global Labour Journal* 2(1), 78–83.

Webster, E. (2010). 'From Critical Sociology to Combat Sport? A Response to Michael Burawoy's "From Polanyi to Pollyanna: The False Optimism of Global Labour Studies" (GLJ 1.2)'. *Global Labour Journal* 1(3), 384–87.

Webster, E., Lambert, R. and Bezuidenhout, A. (2008). *Grounding Globalization: Labour in the Age of Insecurity* (Oxford: Blackwell Publishing).

Workers Rights Consortium (2010). *Preliminary Report on Minimum Wage Violations*. Available at: http://www.workersrights.org/linkeddocs/Bangalore%20Minimum%20Wage_Preliminary%20Report.pdf (accessed 10 October 2011).

Workers Rights Consortium (2010b). *Designated Suppliers Program*. Available at: http://www.workersrights.org/dsp.asp (accessed 20 September 2011).

World Bank (2011). *Lao PDR Investment Climate Assessment – Policies to Promote Growth in the Non Resource Sectors* (Washington, DC: World Bank).

World Bank (2012). *Lao PDR: Labour Standards and Productivity in the Garments Export Sector – A Survey of Managers and Workers* (Report No. 70809-LA) (Washington, DC: World Bank).

World Trade Organization (WTO) (2010a). *Labour Standards: Consensus, Coherence and Controversy*. Available at: http://www.wto.org/English/thewto_e/whatis_e/tif_e/bey5_e.htm (accessed 20 September 2011).

WTO (2010b). *Regional Free Trade Agreements*. Available at: http://www.wto.org/english/tratop_e/region_e/region_e.htm (accessed 14 August 2013).

Yale Law School (2011). *Tearing Apart at the Seams: How Widespread Use of Fixed-Duration Contracts Threatens Cambodian Workers and the Cambodian Garment Industry* (New Haven: Yale University, Allard K. Lowenstein International Human Rights Clinic).

Yimprasert, J. and Candland, C. (2000). *Can Corporate Codes of Conduct Promote Labor Standards? Evidence from the Thai Footwear and Apparel Industries* (Bangkok: Thai Labor Campaign).

Yoon, Y. (2009). *A Comparative Study on Industrial Relations and Collective Bargaining in East Asian Countries*, Dialogue Working Paper No. 8 (Geneva: ILO).

Zhu, S. and Pickles, J. (2013). 'From "Bring-in" to "Go-up", "Go-west", and "Go-out": Upgrading, Regionalisation, and Delocalisation in China's Apparel Production Networks'. *Journal of Contemporary Asia* (forthcoming).

Index

Printed and bound by CPI Group (UK) Ltd, Croydon, CR0 4YY